NEW TECHNOLOGY DEVELOPMENT TREND

JOHN LOK

Made with ♥ on the Notion Press Platform
www.notionpress.com

Contents

Preface

Preface

This book brings readers to enter our future technological world to let you to feel how we will encounter in future predictive technological innovation or invention to influence our daily life. This is future US and UK technological fiction to explain what images we will feel when our societies were become technological life. I write this book aims to explain how future technology change will influence human living of standard. I suppose that human living of standard will become better if future human technology will be improved. Otherwise, human living of standard will become worse if future human technology won't be improved , even will be fallen down. I shall give US and UK technologies development past and present evidences to explain how their technologies development will improve human living of standard. I hope readers can make personal judgement to evaluate whether technology development will improve human living of standard.

This book divides three parts. The first part explains what future technology factors will influence US economy development. The second part explains what future technology factors will influence UK economy development. The final part that I shall give some economy theories to explain why these kinds of technological factors can influence future these both countries' economy growth. This book part one and two concern to be given on my opinions and some economists' opinions to UK and US both countries government and businessmen what will be the most urgent important needs to develop on the technological product investments aspect. I shall indicate the online teaching technology and environment protect technology, such as natural energy, farming environment protect technology etc. as well as automatic technology in manufacturing industry , such as human intelligence products to give my reasons and some economists' reasons to explain to support why these three technological developments will

be valid to invest to develop to UK society in the future.
In my this book part three, I shall attempt to explain how and why ecommerce may be one kind network human job. Also, I shall indicate reasons to explain why human network behavior may bring direct or indirect influences to economy growth or recession in our global societies in macro and micro economy view. Why leisure changing environment may influence human behavior , even economic environment changes. I shall indicate cases to explain any possible human social activities may bring direct or indirect influences to cause our social economic growth or recession in consequency in possible. I hope that my readers can feel more understanding whether what real meaning of behavioral economy is the relationship between our behaviors and our economy.

Prologue

Table of content

How human productive behavior may influence economic development

- New Zealand farmer individual wine productive behavior
- America high technological productive behavior
- China share market investing behavior

Why has any individual country have many people invest share behavior which can influence the country's macro consumption desire?

Can technology influence human shopping behavioral change?
p.180-200

Why and how human behavior may influence the country's economic growth or recession?

Technology how impacts human behavior changing?

How and why employees behaviors may influence economy development?

Robots invention whether they can help organizations to raise efficiencies or inefficiencies?

Why social behavior may influence organizational strategy needs to be changed ?

How and why human behavior may influence economic growth or recession?

CHAPTER ONE

UK Future Technology Development Trend

Online teaching technology

Future, online teaching method will be popular to be applied to teach to any university, even secondary and primary schools. Because internet service is free charge to any students in any countries. Many different age students who can know how to apply internet as well as internet studying is very convenient to any students who can to internet to learn or study in home or public library or school library conveniently. Teachers do not need spend much time to teach students in classroom. They can use internet to teach teachers by face to face seeing and talking to their individual student from every student's computer. So, students do not also often spend much time to go to school to learn. So, developing any fast speed and time saving and talking and listening online teaching methods will be popular needs to any UK primary and secondary and university students in the future. It will be one new technological teaching method to change the traditional classroom educational method in UK and schools. For example, when one UK student who had left UK and is living in another country long time. If any UK school did not provide online teaching service to any UK students. It means that the UK citizen can not choose study himself/herself any UK school if who still hope to study any UK course when who is living in another country. Even one foreign student who does not go to UK to study, if he/she can find any UK primary or secondary or university to study from online. Then, the UK school won't lose one foreign student, due to it does not provide online teaching method to any foreign students. So, online Technology educational learning method will be one popular learning method which is enhanced, supported, mediated or assessed by the use of electronic media. Technology also enhanced learning may involve the use of new or established technology and/or the creation of new learning material. It may be deployed both locally and at a distance (i.e. a combination of traditional and e-learning approaches), to learning that is delivered entirely online. Online learning technology characteristics (features) include identification of a project lead for each area of any learning strategy, identification of two " quick win" for example lecture capture, electronic submission and feedback.

How can online technology enhance learning at UK any schools? It will include these several aspects to analyze. On identifying, prioritizing and innovation hand, online technology is a process for resourcing, prioritizing, acquiring and evaluating school software and hardware for UK any school needs. On staff development learning plan and a student skills development plan hand, UK schools need to establish a base-line policy on the standard (minimum) technology enhanced learning expectation for education each program and module and a mechanism for updating the schools' policies. On evaluation and research hand, a mechanism for engaging the owners of the technology enhanced learning strategy with best practice in the sector including contributing to and benefiting from pedagogical research and the evaluation of the student experience to UK any school.

Thus, UK schools can apply online technology to develop on educational aspect, such as digital literacies and appropriate technical skills that equip UK students for life-long learning, graduate level employment and professional practice, be empowered to learn how to learn with online teaching technology, using online technology to engage in interactive, creative and co-constructed learning with the potential for online learning in an interdisciplinary and international context, using online teaching technology to engage in learning with and from people from anywhere in the world, be supported on placement and in workplace learning through mobile applications and other supportive technologies that facilitate their online learning when away from the classroom, having access to innovative methods of online learning teaching and assessment that are the foundation of a research-lead academic environment, engaging with UK schools in developing , implementing and reviewing the technology enhanced learning strategy. Thus, in the future, it is important to build a capacity to the online education strategy to adopt future learning innovation and student individual online learning need (demand) to UK any school online teaching trend.

There are many examples where UK academics working in isolation or in small UK teaching organizations or classroom learning groups have developed teaching innovation that have a positive impact on UK students' academic experience , but these have remained isolated to particular modules or occasionally program. The aim of education researching online learning process is to identify the good online teaching innovation that is being developed and to prioritize those that have the potential to make a significant contribution to improving the academic student experience at UK any schools. This online teaching process will need any UK schools which can plan how to apply limited resources necessary to achieve online teaching. In addition, the online teaching research process would evaluate and prioritize large scale educational software and hardware requests for primary, secondary and university students' requests. An important part to this process will be to ensure the integration of online educational products and packages that school staff and students regular use to make routine working and access as seamless as possible.

Decisions about school administrative online technologies should not be taken in isolation before assessing the impact on UK teaching staff. In addition, a range of techniques such as, online expert facilitation, coaching and peer support will be used to support individuals, groups or longer academic units, who are learning on major technology enhanced online learning projects. Staff engagement may also facilitated through incorporating technology that is used in teaching staff research and/or professional activity that can be cooperated into their teaching.

Online learning technology can develop UK students skills, UK schools need to understand how UK students understand technology and learn with it, therefore the digital literacy strategy needs to be considered as part of the overall strategy as well as the relevant skills development in UK employability strategy. So, in the future, online learning strategy will make it clear that students will develop technical skills the appropriate level for graduate employability and professional practice. Also, in the future, the online technology can enhance learning working group to discuss external development, that are of educational strategic importance, understanding and evaluating current best practice and research and understanding and evaluating the online educational strategic contribution that pedagogical research and student feedback can have on online educational strategy, policy and practice. The e-learning unit is responsible for informing and educating. This could be done by, for example, providing a short digest of relevant information for each meeting and by setting aside a proportion of each school meeting to discuss a topic of particular online educational strategic interest to every school. Academics that have not got a specialist interest in online educational technology enhanced learning will need relevant information at an appropriate time. This could be provided at a school department or faculty level and this will have clear links to the staff online teaching development plan. Hence, online educational development strategy will influence any UK educational school technological improvement in the future.

Environmental protection technology

Why does environment technology valid to UK businessmen and government to develop? Nowadays, global air and water pollution is serious. Even, UK has many farming is polluted by the water and air pollution. It will influence UK farmers' income if whose farm land (natural resource) is polluted by water or air (natural resource). Even it will influence UK citizen will encounter food shortage if UK farmers can not grow any fresh and health food to provide the enough food numbers to eat every day. Moreover, air and water pollution will influence UK citizen drink the polluted water and breathe the dirty air to live every day. This natural resource (air and water challenge) will influence UK citizen health to cause illness , even death every easily. So, UK government can not neglect the natural environment pollution challenge. The environmental protection technology will help the UK and development countries to solve the challenge of climate change to avoid or reduce farming, foods, or vegetable or fruits or rice, pork, livestock numbers loss threats, i.e. the development and deployment of low carbon energy technology, including technology for the efficient use of energy. The commercialization of low carbon energy and energy efficiency technologies in the UK, with a specific focus on the demonstration and deployment phases of bringing low carbon technologies to UK market.

The UK Government needs to deliver a low carbon economy and to meet UK ambitions emission reduction target. So, low carbon and environmental protection technology researching and development will reduce the carbon intensity of energy production as well as reduce energy demand, towards meeting the contributing UK's ambitions production as well as reduce energy demand, and renewable energy goals. The use of energy (including transportation fuel) and the UK's targets on climate change, for example, by helping the UK make a step change in increasing deployment of renewable energy, improving UK energy efficiency and helping low carbon technologies reach the market. The development of low carbon technologies, and to realize the benefits of doing so in terms ensuring security of energy supply for the UK future economy development.

In UK, private sector investment in technology innovation in the low carbon energy sector will other sectors of the economy. So, in UK energy technologies are likely needed to be developed to avoid dangerous climate change, or an acceptable cost. So, in the future, UK government will need to consider to research environment protection and low carbon energy technology. The activities will reduce carbon emissions, or have the potential to reduce carbon emissions on the longer term, through the use of energy technology will accelerate development and deployment of low-carbon energy and energy efficiency technologies will capacity in the demonstration and deployment of low carbon technologies. Innovation in the energy sector is the only way to identify, develop and reduce the costs of new and improved technologies for the extractions, generation, distribution and use of energy. It has long been an important means of achieving the UK's energy policy aims of a secure and affordable energy supply, as well as to develop the environmentally friendly technologies that are required in UK response to climate change, i.e. nuclear, wind or water, sun energy technology, which is future new energy technology is suitable to research to create to apply instead of current electricity energy.

How global warming influences UK agriculture growth. Scientists have also been fighting the use of chlorine in municipal water systems to kill various strands of bacteria. Chlorine reduces by about 80% the number of alimentary tract diseases relative to polluted, unchlorinated water. A relatively new genetically modified agricultural products. They were partly successful in Europe, such as UK (some countries banned genetically modified products) in spite of the fact that neither history nor research supports their case. People began to modify plants as early as the beginning of the agricultural revolution (8000 to 10,000 years ago), when they started seed selection and who have continued ever since. The green revolution of the 1960 year brought about strains of grans and rice more resistant to a variety of local conditions. The effects have been that countries like India, which had suffered from recurrent famines over the millennia, became self-sufficient in food due to the resultant sharp increase in agricultural productivity.

It was a real science and technology over the poverty dominating most of human history. But it is precisely the products of science and technology that ecologists are so deathly afraid of. In an interesting study in a quarter (28%) of clinically analyzed cases of obsessive compulsive disorder were cases resulting from the fear of global warming.
To destroy the modern, whether industrial or postindustrial, civilization, human have to destroy an important engine of economic growth, that is its energy sources. And this is what eco-warriors try to achieve under the banner of against global warming. Thus, UK government will have responsibility to attempt to research new technology to fight global warming challenge for itself farmer benefits and even global benefits both on the future.

Future Economist global warming technological protection economic influence opinion
Some future economists indicated reasons to explain why UK government and businessmen needed to consider how to develop natural environment protection technology to avoid global warming challenge to influence UK economy development. They indicated the anthropogenic (human-made) global warming resulting from the increase in "greenhouse gas". They offered their perspectives on the scientific valid of anthropogenic global warming phenomenon, its probability of occur and expected consequences and is dominated by technologists, economists and political scientists, who considered the need to make the horribly costly adjustments in energy generation and usage suggested by climate alarmists.
Many stress that global warming is primarily caused by other phenomena than human use of fossil fuels or human activities in general. They are looking at the activities of the sun and impact of the larger universe as the main source of global warming and stress that global warmings (plural) happen intermittently with global cooling. I shall explain why global climate warming will influence to the political and economics of the issue to UK country. For it is the latter, rather than the global warming itself, that will pose

a challenge to the Western world, such as UK and the world at large in the future. Scientists concerned who should move forward with policy measures to avert the alleged disaster. They also apply manufacturing theories to support enough to frighten politicians into action and scare societies into acceptance of measures that would sharply reduce UK citizen their living standards. Otherwise, UK politicians had support that bureaucracies were established, money allocated and lobbies created dependent on the new kind of subsidies. In consequences, climate alarmism and resultant interventions in national economies and human activities have become the increasingly wide spread and increasingly cost reality. With the growing availability of money distributed, and even more promised, a range of benefit of the global warming machinery has been on the increase. So, if UK government did not concern how to innovate new weather protection technology to avoid climate change adverse (poor) influence. It is possible that billions of dollars of UK public money are needed to spend on research global warming challenge because global warming will influence UK agricultural industry. UK agricultural industry is one important export income source to raise UK GDP income every year. If global warming become very serious to influence UK weather to be bad to cause UK farmers who can not grow good taste food and vegetable to supply to domestic and overseas food consumers to eat. Then, UK will loss much GDP income from local agricultural export sale. It seems global warming and agricultural production which has direct relationship to influence UK economy development in the future.

The main problem with climatology is that it must be based as already stressed on very many variables affecting climate and too few hard data necessity. Differences apply not only with respect to the scale of changes obtained, but even to their direction (rising or declining temperature). Some weather scientists indicated to concern global warming challenge. In consequence, it would be impossible to discover if and where errors were made not only in estimating relationships between variables but also in the quality

of data used. (Hauser, J. Tellis, G. J; Griffin, A. 2006) They were comparing average temperatures measured some 30, 40 or 50 years ago by, say, 90 % weather stations in the countryside and 10 % stations in the cities with contemporary average temperatures measured by weather stations located today on 50:50 basis in the countryside and cities. Then, one could obtain the increasing temperatures without any real world climate or even weather changes. Comparability would be ensured if the same number of countryside-located and city-located weather stations had been compared for different periods. The alarmists intentionally mix up " temperature growth" with the trend of temperature growth. To give an example, if in the first decade the temperature grew by 0.5 % degree, in second decade it grew by 0.3 % and in third decade it grew by 0.1%, what was registered was a growth in the temperature, but certainly not a trend of growing temperature. A fourth decade should, on the basis of the trend, bring about no change in the temperature.
To conclude, scientists believed that global warming was caused by human's bad behavior more than natural environment influence. So, it is human's responsibility needs to solve this challenge, due to who feel earning profit aim is more important to protect natural environment, e.g. air and water pollution , due to manufacturing process is the main factor. So, UK has responsibility to attempt to research how to solve global warming challenge , such as it has many famous scientists who can devote their scientific skills to cooperate to solve global warming challenge with other countries' scientists. Some weather scientists also hypothesized that human may be at the end of the present warming period. If they are right, it would be bad for humanity, as warmer periods have always been associated with better conditions for economic activity. To sum up, scientists believed that global warming will influence human economic activity to be bad.

The psychology and political economy of global warming success to UK farmers

Climate alarmists were able to convince a large part of the Western public and a majority of Western politicians of the cause of fighting against the global warming. It supposes itself in an instinctive preference for collectivist solutions in economic and social spheres, with negative to disastrous consequences when scientists are applied in practice, so UK government needs to concern global warming challenge, due to it is possible that it will influence UK natural environment weather to be poor to influence many UK farmers‘ agricultural and vegetable and fruit and rice wheat etc. food growth successfully. What is the global warming influence to cause disease? For example, ecological alarmists and activists (eco-warriors) never admit they are wrong, they long pursued their fear mongering campaign against chlorine. Their success in branding DDT a dangerous substance had a negative impact on the malaria eradication campaign in poorer parts of the world. Alternatives to be have been far less effective and the result has been the resurgence of malaria cases and the manifold increase in malaria -caused deaths to the largest extent in Africa.

Automation technology in manufacturing industry

Nowadays, UK computer and space explore technology had reached the mature stage. It means that UK government ought not need to continue spend much resource to research these two kind technologies. Otherwise, the automatic manufacturing technology, e.g. human intelligence new product. It has need to develop because human intelligence machines will bring beneficial to satisfy human everyday life need, e.g. hospital patients' activities need, if the patent who can not walk easily, but the human intelligence machine can assist the patient walk to anywhere conveniently. So, he/she does not need to sit on wheel chair and apply the human intelligence machine man to help him/her to drive on the intelligence automatic driving vehicle to go to anywhere conveniently.

Otherwise, increased automation in low wage countries, e.g. China, Korea, Africa, Hong Kong etc. which have traditionally manufacturing firms, could use automatic technological manufacturing to bring lose cost advantage and potentially lose their ability of achieving rapid economy growth by shifting workers to factory jobs. So, UK government and businessmen needs to consider automation technology development, i.e. 3D printing manufacturing industry will encourage UK companies to move manufacturing process, closer to gain the biggest advantage from this 3D automation technology development.

A growing concern of premature de-industrialization in energy and developing countries could require new models and a need un-skillful the UK workforce. In the future, the best way toward for UK cities will reduce their exposure to automation is to boost their technological dynamic and attract more UK skilled workers. Automation technology progress can give UK manufacturers' employee benefits, such as long term healthy productivity improvement, raising productivity efficiency and product quality, macroeconomic and microeconomic effects of automation technological change, it's change will be beneficial to UK society, i.e. automation active labor market policies, which could help UK job seekers find jobs from training to incentive to support self-employment to create high technological job employment chance in UK society. So, raising science, technology, engineering and math subjects update skills level are needed to UK any universities, which can be increasingly important in UK society, these factors could complicate the ability of UK high automation technology education to adopt to the UK automation manufacturing technological change. A talent mismatch already exists in UK, with many well UK educated workers can find employment in lower-skilled jobs. To combat this, greater coordination will be needed between the education, training and employment sectors in UK society.

Why are high automatic technology product development models needed to research to UK any manufacturers? UK government and manufacturers need to consider how to achieve high technology product development models. According to Hauser et al. (2006) indicated the high technology (high tech.) development process, is influenced by the innovative process, bringing products on exception value which stimulate product market demand. Innovation provides products the specific basis for which world economies compete with each other on the global market. Able to find new solutions, innovations generate significant changes in existing markets, destroy them, or create new marketing (Hauser et al. 2006). So, UK manufacturers need to concern on any manufacturing high technology product development process because which can influence any new products development to manufacture to sell to any overseas or domestic both markets successfully.

What is high tech. product meaning? Mohr et al. (2010) argues that there are two reasons why it is important to clarify and specific high technology : (1) due to the impact of technologies on the economy, attempts are made to classify economic production and incomes ; (2) due to the impact of high tech. on the environment. Standard marketing strategies are being modified and adopted , therefore, it is necessary to know the products to focus on. Why UK manufacturers need to consider high technological product process. Nowadays, high tech. products are complex, advanced, requiring specific technical knowledge, which is technologically not discontinued and being produced at the companies which have twice as many technical personnel and invest twice as many in scientific research and development than other companies. Moreover, these products are time-sensitive as scientists are continuously searching for new approaches for invention of more advanced technologies which make all preceding ones lower-ranking. The most important, nowadays global consumers will adopt the particular technology. It means that global customers may delay adopting new high-tech. products and in order to mitigate the prolonged uncertainty require a high degree of education and information about the product and need post-purchase reassurance. Anyway, nowadays customer individual needs in high tech. environments are characterized by sudden changes related to unpredictable fashion. Even, consumers concern about how to preserve new product' competitive technological standard is completely incompatible with technological uncertainty. The most important factor is the prevalence rate of any new products development process, which is influenced by slower than of traditional products. In many cases high-tech. automatic product market are being materialized slower than which are expected. The technological uncertainty challenges will exist in development process, such as uncertainty related to the timetable for development of the question whether the new product will be function as promised. In automatic high-tech. industries, the time requires for product development is difficult to predict as , commonly, it takes longer than expected , uncertainty related to unanticipated consequences and uncertainty about the product life cycle related to competition products. In conclusion, these factors will influence new automatic technology product development process unsuccessful, so UK manufacturers will need to concern on any high technological automatic product's manufacturing process.

Future economists predict automatic technology how to influence future UK economy

Before, all over the world presented picture of demonstrate in London on the occasion of the meeting of the G20. Some economists indicated disastrous economy consequences will occur to any one of Western country , such as UK, so if any one of Western country did not consider automatic technology development to itself country. They indicated one example, such as material incentives to produce disappeared throughout Russia and, when Society leadership called off the experiment, the country faced industrial output reduced to 10% of what had been registered in 1914 and agricultural output reduced to such low levels as to cause widespread famine.

Why would UK encounter disastrous economy consequences if UK government did not encourage manufacturers spend money to invest to innovate automatic technology industry? According to a variety of anthropological studies, a collectivity is unable to operate efficiently with everybody giving talent workers have chance to devote whose best effort to manufacture any high technological products, e.g. human intelligence vehicle or airplane. Hence, economic incentives are needed to UK manufacturers to invest high technological automatic industry development. Because the economists predict UK will have many talent worker numbers, their number will be more than a certain number of normal effort workers, due to UK technological education level is very excellent to provide to train many young technological manufacturing students to find this kind of high technological manufacturing job. So, the high technological manufacturing job seekers will increase and it won't decrease to UK job market in the future.

Assuming that UK high technological automatic manufacturing workers who would desire only to introduce changes in the workings of the international economic order and policies of countries participating in the present economic order rather than change the order itself, what will be UK manufacturers their specific economic preferences in the future? It implies tnat either concentrate on spending more investment to automatic high technological development, e.g. human intelligence automatic high technological products or still concentrate on spending more investment to common traditional technological products.

However, UK was a developed Western country which had had strong automatic high technological development effort very long time. Otherwise, it compared to some developing countries, such as Asian China, Hong Kong, Korea etc. Asian countries their future economic growth rate will show un- surprising , different patterns, so the Asian countries has weak effort to invest high automatic technological product development, such as human intelligence technological development. The catching-up process suggests low economic growth rate in the high automatic technological product development to the Asian developing countries in the future.

Hence, the future economists predict that it views as probable successors of the Western world economic leadership if any Western country , such as UK manufacturers who prefer to invest to any high automatic technological products development , e.g. developing on human intelligence automatic technological products more than traditional common technological products development. On the one side, but it seems important to stress that two very poor countries among the challengers-China and India-are examples of countries that changed their institutions and economic policies from no or little economic freedom to more economic freedom. Because there two countries whose governments prefer to lend loans to encourage their country manufacturers prefer to invest high automatic technological products manufacturing. On the other side, attitudes toward foreign direct investment (FDI) have undergone change since the 1960 s and a large majority of less developed countries, e.g. China and India are now competing strongly among themselves and with developed market economies for direct investment from multinational companies. So, UK will face China and India high automatic technological product competitors in the future. And in fact, all countries that joined Western developed economies did that without much (if any) external inflow of public resources. It is right time that UK government needs to lend loans to encourage domestic manufacturers to invest high automatic technological products to raise whose international high technological products sale effort to win its future competitors. So, machine resources will be increased demand to o UK manufacturers if who chose to spend machine resources to innovate to manufacture any new and high technological automatic products to raise human daily life needs in the future. It means that it is right time UK manufacturers need buy much machines to prepare to manufacture many future high technological automatic products when these machine prices are low. Because the future global machine prices will possible be raised if many China and India manufacturers will also buy many machines in the future. For example, USA government had provided much financial support to assist sugar cane producers to develop their businesses. And they are dependent to a much larger extent than sugar cane producers and sugar processors in the USA on government. Without very high subsidies to renewable energy generation, they would not have survived at all. So, USA government had been the first country which could lent much financial assistance to encourage domestic renewable energy generation manufacturers to develop high technological energy manufacturing business. So, UK government needs follow USA to lend financial assistance to encourage domestic high technological automatic industry development.

Future economists also predict China and India will be competitors for future leadership in the global economy, special high technological products. China has been the media and analyst's favorite for quite some time. Quantitative projections have seemingly supported such expectation. Such as China and India had manufactured many high technological new space rockets products, ocean war large ships etc. Moreover, China has become one of the major world trade players in the early twenty-first century.

Many long-term forecasts, assuming similarly high economic growth rates in the decades ahead, predict that China will surpass the USA in terms of aggregate GDP somewhere between 2020 and 2030 or later, say between 2030 and 2050 year. The future economists conclude on the basis of these predictions that China will not only pass the USA in aggregate product (GDP), but its economy and economic policies will influence the rest of the world to a similar extent that the USA does at present.

I stressed a very important point, namely that the UK future high technological automatic product competitor China and India, namely that economies not only grow, but in the process change their structure. China and India have been industry very rapidly (the first transition) and building the physical infrastructure that accompanies industrialization changes to technology in the future. However, at a certain per capita GNP level the two countries, such as China and India will face another structural shift when which technological development will reach the mature stage in the future. China and India had been primarily historical pattern of economic development because the shift in the role of engine of growth from industry to services is to a much greater extent a qualitative shift. Both higher and different skills are required. And, even more importantly, interactions generating ideas driving the highly human-capital-intensive service economy require a much freer environment, not only in the economic area. Chinese exports have been heavily labor-intensive. This being the case, they contributed to the expansion of industrial employment, offering for the first time in the history of China a taste of (very modest) prosperity to more than 100 million new industrial workers and their families. This is the major component of the success accomplished by Chinese economic growth. Richer trade partners create room for more trade, so the Chinese should hope that intra-South trade, that is, trade between the emerging economies of Asia, the Middle East, Africa and Latin America, will open up new and growing opportunities. I presume that if Western economy , such as UK did not developed high technological automatic industry to stable their social welfare, so thoroughly slowed down their economic growth.

Will it allow China to accomplish the transition to a mature, innovation, service-sector-based market economy? It has allowed the economy to industrialize much more successfully, even if the labor shift from agriculture to industry has not yet been completed. But it is a long way off the next major test: the second high technological industry transition of the economic structure to China. Bear in mind that Russia attempted it twice and failed at both attempts.

But even, assuming that China at some point in the future does succeed in accomplishing the second transition, will it be able to supersede the USA, for example, as the main global high automatic technological innovation center if it wants to become the No.1 global high technological industry economy? Given the nature of the centralized state and its stability to collect financial resources , China's ability to increase research and development expenditure to high automatic technological products and to hire a mass of researchers, engineers, technicians and other specialists should not be doubted. This process in already taking place.

But , again, Soviet Russia already exceed the USA in the R&D/ GDP ratio in the 1970s, long before the communist collapse, with no effects on its innovativeness. Inputs matter less than outputs, quantity in the innovation process mean much less than quality. The latter characteristics depends importantly on economic, civic and even political institutions. Otherwise, independent India had three options open to it in 1946s. It could pursue spontaneous economic development, with some state intervention to be sure, along the lines of basically free market capitalism; it could turn the clock back and try to recreate the rural-agricultural and handicraft based. The dominant way of thinking was Society -style priority to industrialization and , within industrialization , priority to heavy industry. In other words, not textiles and clothing, which has been developing well in India since the mid- nine teen century, but production of sewing machines and , even better, production of machines the produce sewing machines.

The results were only to be expected. The heavy stress on the expansion of capital-intensive heavy industries in a very poor country quickly strained the ability of the Indian economy to generate adequate savings. Moreover, some of these industries were above the level of industrial competence of an underdeveloped economy. Thus, the amount of required resources (capital, skilled labor) was usually larger per unit of output than in the same industries in more mature, richer industries economies. In another view point, India will develop light industries, just as any other poor country with a great deal of unskilled labor, had a comparative advantage and no less importantly, an economy in which, due to their low capital/labor ratio, light industries could employ many more people, spreading prosperity more widely in a poor country. So, it explain that why China will have more effort to develop heavy high technological industry in the future. Thus, India got less economic efficiency, less employment than in a spontaneously developing economy, less ability to compete internationally in light industries suitable for an underdeveloped economy and finally got heavy industry unable to compete even on the domestic market and, therefore requiring no less heavy a dose of protection. Overall India got an underperforming economy, in particular in its relations with the rest of the world.

To conclude by comparing the performance of the traditional sectors of the Indian economy and the performance of its modern, human -capital-intensive subsector of manufacturing and skill intensive service sector. The latter both employ workers with high- and medium -high skillful level (in branches ranging from computer software and biotechnology and pharmaceutical high technological light industry). India is ahead of China in terms of the output and export of such products and services. Thus, it implies that UK ought concentrate on developing high automatic heavy high technological industry, e.g. human intelligence technological products because these industry is not better development to other many countries' strong effort , such China and India large population countries.

CHAPTER TWO

US future Technology Development Trend

Developing countries cities technological competitive investment factor

Some economists predict developing countries every city in the global 750 is projected to have a larger future technological economy growth. But the diversity of developing countries' economic performance is large. Developing economy cities, such as China, Japan, Hong Kong, Korea cities can grow rapidly by acquiring capital and technological know-how and putting them to use by their rapidly growing urban labor forces. Even, these developing countries cities' rapid technological development can impact to US labor market supply.

Due to future rapid technological development to these developing countries, the result will cause developing countries cities, such as Asia China, Hong Kong, India cities economy growth will rapidly. Otherwise, developed western countries cities, such as US, UK, Canada, Australia, Span, Germany etc. countries lie close to the technological frontier have stable urban populations and more limited investment and job creation opportunities. It will influence these developed countries' economic growth is slow than the developing countries. Due to developing Asia countries cities will prefer to invest more technological development to compare to developed countries, such as US, UK, Canada, Australia, Span,

Germany etc. countries. Therefore developed countries, such as US, UK, Canada, Australia, Span, Germany etc. countries tend to grow more slowly. It seems developing Asia countries, such as Hong Kong, China, Korea etc. countries urban and central cities economic performance within developing countries will be better than developed countries, such as UK, US urban and central cities within five years.

Thus, I suppose that future developed countries, such as US the speed of technological development will be slower to compare developing countries, such as Hong Kong , China, Korea, India etc. countries. Then, it will bring this question: How to solve developed countries, such as US, Washington, New York etc. cities future economic slow growth challenge? For future developed country cities, such as New York, Washington etc. large cities' technological investment and location decision, it will need understand that diversity is essential. Various factors can have an impact on US country intra-national urban cities performance, including sector structures, agglomeration benefits, infrastructure quality. For example, US country central government needs have tolerance of diverse performance, land supply and US city governance plan. It aims to raise US central and urban cities' future economic performance. Moreover, US country will need to concern above issues to arrange how to improve central and urban city economic development . Specially, it will need to concern technology innovation to encourage many overseas technology investors to anticipate different kinds of technology products innovation, e.g. human intelligence machine, mobile, computer etc. high technology products. It aims to absorb different countries' new technological knowledge of different kinds of high technological products to excite US domestic or foreign countries consumers' desires to choose to buy many different kinds of high technological products to raise US GDP growth in US technological product sector industry in the future.

What are the trends impact on rural
America's future economy?

Some economists indicate that there are five trends reshape to impact rural America' future economy. They include that digital economy will shift future America rural economy. US quality of life will change a lot, the US rural economy will stay uneven, US commodities will compete in global markets and will give less benefit to US rural economy and US new products will revolutionize US agriculture economy.

The first aspect impacts to US agriculture economy, the US future rural economy stays uneven. Growth will concentrate in 4 out of 10 rural places and they have scenery, a retail hub, or one next to a city in US. The impact on rural America includes some rural places will try to manage growth , but many places on a quest for new economic engines. Thus, it will bring these questions to US rural economy impact, such as : Who will be US businessmen clients? The struggling farmers? The struggling farm-dependent country? The booming mountain area? The rural area transforming into city? The second aspect impacts to US agriculture economy, due to US commodities will compete in global markets if it will bring a smaller benefit in US rural economy. Then, US farm scale will cut costs and competition fewer US farms and places will depend on farm income. What is the impact of commoditization on rural America? There are more farms depend on area jobs and it creates a new imperative to add value. What is the impact of commodity on US clients? The need for competitive commodities remains, but the payoff for added value is rising and community impacts are important.

The third aspect impact to US agriculture economy, new products will revolutionize agriculture. Major, shift from commodities to products, spurred by biotech, means two agriculture in the future and two rural America . Hence, US future agriculture determination the rural economy will be declined. A future new US agriculture supply chain integrator will be caused from the traditional farming supply chain procedure the change to outsourced contractor farming supply chain procedure, such as: In beginning, from farmer will outsource supply chain contract to processor, then contract to distributor and contract to food retailer final step.

The fourth aspect impacts to US agriculture economy, what will be two agricultures impact to the US future economy? The first US agriculture impact will be US commodity agriculture. It focuses on production capabilities, farming foods production will be thin margins maintained with technology and big sale. The second US agriculture impact will be product agriculture. It focuses on consumer needs, farming foods production margins will be protected by capturing value and building business relationship. Thus, it will bring these questions concern what the impact of product agriculture which can influence to US economy. How to apply biotechnological agricultural techniques and farming product application to raise US farming productivity? How to build world class agricultural (farm) producing chains that benefit to US farming produces production method? How can US faming foods producers participation or lead in product chains?

In the fifth aspect impacts to US agricultural economy, it is digital economy impacts US rural agriculture. Future diversity economic base can encourage US farming product agriculture to enter rural digital agricultural service consumption sector to change US agricultural consumer individual shopping habit. What will impact to US farmers? Digital agricultural economy will bring these questions: Who will lead in broadband digital agricultural technology? How to launch e-markets and businesses and how to cooperate with US rural farmers? What will impact on US agricultural on natural resource management aspect and on how to US commodity to build amenity strategies? However, rural America's future will be shaped by policies that will encourage agricultural technology adoption, it will enhance farming worker skills, it will improve rural quality of life and it will ensure access to capital to bring US agricultural economy growth.

High level education and high
birth rate factor influence US technological economic growth

According to John G. (2016) indicated that " a point forecast is that GDP per capita will rise well under 1% per year in the longer run, with overall GDP growth of a little over 1 to 12%. The main drivers of slow growth are educational attainment and demographics. First, rising educational attainment will add less to productivity growth than it did historically. Second because of the aging and retirement of baby boomers, employment will rise more slowly than population which in turn, is projected to rise slowly relative to history." Thus, it seems that some economists believe education and born rate will influence future US employment method. They assume that the employment and growth ratio will rise and unemployment ratio will deadline of US will increase birth ratio and future there are many young people (students) can have chance to accept high education degree to raise whose education level to prepare to enter different kinds of high educational jobs market to work in the future.

Considering a more growth accounting perceptive on productivity growth. According to the analysis in Fernald (2015) which uses a multi-sector growth model for the projections. It indicated that although, the details differ and if bases projections on TFP data since 2004, it also implies a preferred point estimate of 1.6% per year in US over this short period. The " fundamentals" of labor-productivity growth (namely, growth in total factor productivity , TFP) have exceeded the actual realization for reasons that reflect the unwinding of dynamic of labor quality and capital depending associated with the great recession. Thus, how to raise labor quality which concerns how to raise educational level to young people in any country will be one labor economic challenge in US in the future.

Fernald (2015) also explained " the shortfall" in productivity growth relative to fundamentals reflects, at least in part, the unwinding of two dynamics associated with the great recession. He indicated that first, at the end of the great recession business had a lot of capital relative to labor, which has attended the need of add capacity to meet demand in recent years. In contrast to the outsized growth in capital deepening from 2007 to 2010 year, we have seen capital "shallowing" since 2010 year. Second, businesses fired low-skilled workers during the recession, which raised labor quality in 2007 year to 2010 year period. As these potential workers have been rehired, the growth rate of labor quality has added less.

Thus, on the one hand, when one country encounters economic recession, many employers will like to employ high educational level and high skilled labor to raise productivity. If US encountered economic recession, but it had none (lacked) enough high educational level and high skilled knowledgeable workers to be supplied to the labor market in US. Then, US will encounter long term economic recession period, it can't shorten economic recession period. Thus, US will need to let many US young people have effort to study to prepare enough high educational level knowledge to do different kinds of professional or high knowledgeable or high technological jobs in future US labor market.

However, US high knowledgeable labor can also assist US businessmen to raise productivity growth. Due to labor quality could be grown by high educational and increasing number of knowledgeable workers. On the other hand, moreover, if there are many US young married parents who will either choose not born any babies for next generation or choose born only one to four children. Then, future US will have a small numbers of young people to be supplied to primary schools, high schools, even universities to study. It also means the university level graduate student numbers will decrease also. So, low birth rate factor can also influence the high knowledgeable labor numbers to be supplied to future US labor market.

So, I recommend that US government needs to encourage many young parents choose born next generation and provides student loans to lend to poor families to support whose children can have more chance to attempt to go to high schools or universities to study in the future.

Socio-economic and political factor impact future US strategic policy change?

US development practitioners are increasingly aware of the role that US social and political structures play in future shaping US's development paths and results. In this context, US macro social analysis needs to understand the ways in which power relations act to circumscribe the opportunities available to poor US people to improve their situation. For example, US donor organizations need to understand of relevant US social structures, such as informal institutions or other relevant US social practices in US.

This provides an entry point for understanding the broader US political environment or challenges in a particular sector or process. Furthermore, any US donor organizations need to place greater emphasis on the analysis of livelihoods and economic opportunities and their relationship to reduce the unequal of gap between rich and poor US citizen in US societies . Thus, US government needs to encourage donor organization participants choose to do the reasonable donor behavioral to aim to reduce the unfair donor spending to the unneeded donor assistance beneficiaries. To let us socio-economic has more balance chance to every US poor citizen in US society. So, it can also raise average every US poor citizen feels better quality of life in US society.

How does the rise of US exports to East Asia factor influence US economy change? Export have become an increasing important source of revenue for both national and regional forms in the United States. How does the primary growth market for US exports to influence US economy growth? I recommend the developing nations in East Asia will the developing nations in East Asia will soon rival today's industrial nations as the most important US trading partner . In view of the rapid growth of US exports and their geographic shift toward developing nations.

Some economists indicates that two tends in recent US export performance are particularly notable. First, US exports have increased rapidly relative to total US output, development with important implications for the entire economy. Approximately 130,000 US firms employing over 10 million domestic workers, export their products (US Bureau of the census,1993). Second,

the geographic distribution of US exports has been shifting dramatically. The industrialized nations of the organization for economic cooperation and development (OCED) account for about 57% of all US exports . However, in recent years of the start of US exports to major trading partners in East Asia. America has been rising and now accounts for 26% of the total. Thus, economic policymakers can no longer make trade related decisions without considering their effect on US exports to East Asia market.
From 1981 to 1987 year, two factors temporarily stemmed the rising tide of 1980 year. First, the US dollar began to appreciate significantly in 1981 year as the United States enacted a restrictive monetary policy and an expansion fiscal policy. From 1981 to 1985 years, the dollar rose almost 50%, thus making US exports more expensive relative to foreign products (Hakki & Whittakersand J, Gregg Whittaker, 1985. " The US dollar recent developments, outlook, and policy options". Federal reserve bank of Kansas city, economic review, Sept./Oct., pp.3-15.) Thus, future US will need to make good international export and import trade relationship with Asia countries, e.g. China, Hong Kong, Japan etc . countries to raise GDP income.

Education factor influences US future
technology labor market change?

In the past, there fundamental forces have shaped US work labor market, includes an increase in the returns to education. General education upgrading and the large numbers of female need to work. How educational attainment, demographics and human capital will be predicted to influence US future labor market. Some economists believe education is useful to influence US future labor market. They indicate the social returns to education policies today depend on the relative prices that labor of different educational levels will command in future US labor market; current US labor market trends appear to leave a large group behind; less educated males. The rationale for social policies target specifically to this population is strengthened if predicted future outcomes in US labor market will lead this less educated male group numbers to fall down.

How to supply educational level components to influence future changes in US labor supplied? It can change in the size of the US working age population, it can change in hours worked conditional on being of working age, and it can change in the skills (effective human capital units) to US workers of different education levels, gender and age . In US , the labor is supplied by both highly educated men and women increased substantially relative to the supply of labor by the less educated. Among US males this is largely , due to an increase in educational attainment; highly educated US males did not differentially increase their supplied compared to less educated males nor did their experience differential increases in their human capital policy in US. For US females, some economists found large increases in the labor supplied by US working age women that are due to both large increases in hours worked and increase in educational attainment. So, I assume that future US will have many high educational level female labors to work in US society.

Today, human history is at the beginning of a growth industrial revolution. Developments in genetics, artificial intelligence, robotics, nanotechnology, 3D printing and biotechnology will be popular to influence US future job market change. For example, smart systems, product-homes, factories, farms or cities with help to solve problems ranging from supply chain management to climate change. The rise of the sharing economy will allow US human to monetize everything from their empty house to their car.

Due to the future patterns of consumption will change to trend high technology life enjoyment, it will cause production and employment will also change to employ high technological production labor. As entire industries adjust, most US occupations are changed to high technological manufacturing industries. When some US jobs are threatened by others grow through a change in the skill sets required to do them. a key element in understanding how the benefits and burdens of the growth.

Nowadays, the current technological changes of humans and machines , but rather an opportunity for work to truly become a channel though which human recognize full potential. As US is a high technological developed country. I assume many US employers will choose to act to be the first high technological and invention manufacturing leader to encourage which labors need to learn high technological production methods to prepare manufacture any new technological products to sell in the future US domestic or foreign both markets. Thus, it is possible that future high technological manufacturing labor numbers will be shared large go e.g. 50 to 60% future total US labor market.

How can future US labor market affect job creation and productivity growth? US future economic growth requires factor reallocation across US firms and continuous replacement of technologies. US labor market influences US economy dynamism by their impact on the supply of a key factor, skilled workers to US new and expanding firms, US growth -favoring labor market includes portable pension plans and health insurance united to the current US employers, individualized wage-setting and US public income insurance systems that encourage mobility and risk taking.

US future economic growth arises as production shifts from less to more successful firms though the reallocation of factors of production . US labor market can advance restructuring. Overly regulations tend to create a system in which a large share of economic activity occurs in US small firms without the ability to grow. US labor market should be organized to promote potential high growth US firms, especially through decentralized and individualized wage setting, portable jobs.

In the future, US will have many key importance of high growth firms. US capitalism entails a process of creative destination. New ideas continuously challenges act structures, giving rise to structural transformation as successful innovations and new products firms and industries will arise and obsolete ones will decline in US society.

Martin , J. P. (2012) studies pointed to high-growth forms (

sometimes known as gazelles) as the main drivers of this process. In the US , an estimated 1% of firms creation 40% of all new jobs and 5% create almost 70% of new jobs. A review of the studies of US firms growth reveals some common findings . US high-growth firms are crucial to net job growth, generating a large share of all net jobs . This is particularly pronounced in recessions, when US high-growth firms continue to grow when other US firms deadline. US small firms are over-represented among high-growth firms, but these US firms come in all sizes . A small subgroup of large high growth firms are major job creators. Such as US high -growth firms are younger on average, US young and small high high-growth firms grows, not through mergers and acquisition and make a larger contribution to net employment growth than do US larger and older higher growth firms, high growth firms are present in all industries. Through they are slightly overrepresented in service industries in US.

Some economists predict that future US will be a flexible labor market, the marginal product of labor and the average wage in an industry should tend toward equally across US firms. Taking advantage of a legislative change to raise cost to US employers a study measured the gap between the marginal product of labor and the average wage in an industry before and after the reform. The gap increased after the legislation, which suggests that the legislation reduced allocative efficiency. Their studies have suggested that total factor productivity could increase by as much as 30% in China and India of they were to attain the US level of allocative efficiency across firms within individual industries. The result implies that plants with low total factor productivity are too large and plant with high total factor productivity are too small relative to the US benchmark of allocative efficiency.

What is allocative efficiency meaning? It occurs when the mix of product produced matches consumer preferences (where marginal benefit equals marginal cost). There products and services are the most profitable, thereby promoting economic growth other research also indicate a strong quantitative effects of strict employment protection legislation on the rate of reallocation in US industries experiments. By relaxing employment protection rules to US developed countries, such as UK, UK etc. with the strictest legislation could increases their reallocation rate by an estimated 50% in the most dynamic sectors, those that benefit most from flexibility.

The effect appears to be particularly strong on the entry-exist margin, which is arguably, especially importation for creative destruction. In future US, if manufacturing industry can have high technological production method to achieve reallocation rate to efficiency to every manufacturing industry. It can have benefit to economic growth, due manufacturing process can be move efficient to avoid cost. Also high technological manufacturing reallocation rate innovation method can influence the future US number of jobs lost in contracting or existing US firms as well as the number of jobs gained in new or expanding US firms in future a certain period divided by the average number of existing jobs. So, US high technological manufacturing reallocation rate innovation method will bring disadvantages to cause many contracting or existing firms job lost, but it will also increase the number of jobs gained in new or expanding firms in US. Thus, if future US manufacturing industry can have high technological reallocation rate innovated successfully. It will raise many new job chances in new or expanding firms in US, but it also have chance to cause many contracting or existing firms job lost or the same time.

What will expected to impact of future computerization on US labor market outcomes? Some economists estimated, about 47% of total US employment is at risk, due to that US wages and educational attainment will exhibit a strong negative relationship with an occupation's probability of computerization. They indicated that the poor performance of global labor markets across advanced economies has intensified the debate about technological unemployment among economists more recently. Indeed, over the past, decades, computers have substituted for a number of jobs, including the functions of bookkeepers, cashier and telephone operators (Bresnahan, 1999, MGI, 2013).

Although the extent of these developments remains to be seen, estimates by MGI (2013) suggests that sophisticated algorithms could substitute for approximately 140 million full time knowledge workers world wise. Hence, when technological process throughout economic history has largely been confined to the machine of manual tasks, requiring physical labor, technological progress in the twenty-first century can be expected to contribute to a wide range of cognitive tasks, which until have largely remained a human domain. Of course, may occupations being affected by these developments act still far from fully computerization, meaning that the computerization of some tasks will simply free-up time for human labor to perform other tasks. Nonetheless, the trend is clear: computers increasingly challenge human labor in a wide range cognitive tasks (Brynjolfsson and Mc Afee, 2011).

Thus, it seems that future US computerization job trend will influence some US traditional labor hand made manufacturing industry to divide some part job duty to let computerization work. However, it will not cause many US manufacturing labor to be dismissed. It can still keep some US manufacturing labors to US employers' needs in the future.

Popular science, technology , engineering and mathematics kind of labor supply will be increased demand in US.

Science, technology , engineering and mathematics workers will drive US innovation and competitiveness by new ideas, new companies and new industries in US. However, US employers frequently concern the supply and availability of this kind of workers. Over the past 10 years, growth in this kind of jobs was three times as fast as growth to general jobs in US. This kind of US workers are also less likely to experience joblessness than other kinds of workers. In the future, science, technology, engineering and mathematics workers will play a key role to grow and raise stability of the US economy.

Bureau of labor statistics, ESA calculation 2010 and 2018 year. indicated these kinds of technology, science and engineering workers had been increasing 7.9% from 2000 to 2010 year growth as well as these kinds of workers had been increasing 17% from 2008 to 2018 year growth. Otherwise, other kinds of workers had been increasing 2.6% from 2000 to 2020 year growth as well as these kinds of workers had been increasing 9.8% from 2008 to 2018 year growth. Hence, these kinds of science, technology and engineering workers increasing rate and increasing level are more than other kinds of workers both twenty years.

The other occupations include positions, such as educators, managers, technicians, health-care professionals or social scientists. The science occupation divides four categories: computer and mathematics, engineering and surveying, physical and life science four categories. The reasons why these kinds of jobs will be trend popularly. We define these kinds of degree holders as persons whose primary or secondary undergraduate major was in a science, engineering or technological field. To using similar logic to what we used in our occupation selection, we exclude business, healthcare, and social science majors.

The US department of commerce, economics and statistics administration analysis showed that a science, technology, engineering and math. (STEM) degree is the typical path to a job related to those kinds of degrees or more than two-thirds of the 4.7 million (STEM) workers with a college degree has an undergraduate (STEM) degree. However, this does not necessarily mean that (STEM) field or their jobs . For example, only 35 per cent of college educated computer and mathematic workers have a degree in computer science or math. Thus, these past data explained that it is possible US science, technology ,engineering and mathematics employers will increase needs and why many US students choose to study these subjects in US. Thus, future US economic growth will depend on developing these kinds of science, technology ,engineering and mathematics industries.

Increase development in genetics, human intelligence, robotics, nanotechnology, 3D printing and biotechnology technological industry

In US future, these kinds of jobs will be needed to increase development in genetics, human intelligence, robotics, nanotechnology, 3D printing and biotechnology. For example, smart systems homes, factories, farms grids or cities will help tackle problems ranging from supply chain management to climate change. The rise of US economy growth will allow US people to monetize everything from their empty house to their car in US. These new technological products development will change US patterns of consumption, production and employment adaption are also be changed by US corporations, US government and individuals.

Why will the technological revolution be broader socio-economic, geopolitical and demographic drivers of change to influence future US social economic and consumption pattern change? Future US most occupations will also be changed. When some traditional old jobs are threatened by redundancy and other new technological jobs will grow rapidly, existing jobs are also changed in the skill sets required to do them. The debate is between some economists foresee limitless new job opportunities and foresee massive dislocation of US jobs. In fact, the reality is highly specific to future US high technological production industry, region and high technological occupation in question as well as how US production workers can be raised themselves ability to actions the upgrade level of high technological production ability from various stakeholders to manage high technological production method change.

Overall, this is a modestly positive outlook of US high technological production employment across future most high technological production industries with jobs growth expected in several sectors. However, it is also clear that this need for more talent in certain job categories is accompanied by high skills instability across all job categories. Combined together, future US net job growth and skills instability result in most US businesses with face major recruitment challenges and talent shortages, a pattern already evident in the result and set to get worse over next five years in possible.

The question is how US businesses, government and individuals will react to these new technological job changes, due to talent shortage, mass unemployment and growing inequality challenges will encounter in future US society.

The current technological revolution does not need become a race between humans and machines , but rather an opportunity for work to truly become a channel through which US people recognize their potential. So, if US traditional low manufacturing skillful workers lack talent to learn new skills to prepare to do future new technological manufacturing jobs, such as 3 D printing, robotics, nanotechnology, biotechnological high technological products manufacturing jobs. Then, it will cause increasing of unemployment rate to some not talent US low manufacturing skillful workers. So, US government or high technological product industry employers need to consider this future unemployment challenge will be caused by high technological products manufacturing changing influences. It seems high technological development will cause these low manufacturing skillful workers unemployed rising numbers as well as high manufacturing skillful workers human capital shortage global challenges will exist.

In the future, the driver of changes to influence US demographic and socio-economic growth. They may include: changing work environments and flexible working arrangements. It means new technologies are enabling workplace innovations , such as remote working, co-working spaces and teleconferencing. Rising of the middle class in Asia markets. It means the world's economic center is shifting towards the Asia developing countries.

Some economists predict that Asia will be projected to account for 66% of the global middle class and for 59% of middle class consumption by 2030 year. In addition, climate change, natural resource will be constraints to a greener economy. It means that climate change is a major driver of innovation as organizations search for measures to help adjust to its effects. As global economic growth consumers are needed to lead to demand for natural resources and raw materials, over explanation implies higher extraction most and degradation ecosystem and these challenges will also impact US employment changes needs. All US government also needs to concern future global economic change influence.

CHAPTER THREE

US Entrepreneurship Innovation How Influence US Future Geography

Some economists believed the interplay between entrepreneur , innovation and economic geography growth in the United states of America have close relationship . Because future innovation is the driving force of growth in the knowledge economy to US . They assumed that if future US entrepreneurships chose to act as new firm formation, which will prefer to accept any new technological innovation to manufacture any new technological products and reallocate any new resources to apply to concentrate on manufacturing any new technological products within firms. For example, Bernard, et al. (2006) find that one third of the net increase in real U.S. manufacturing output between 1972 and 1997 is due to the net adding and dropping of products by surviving firms, a contribution that dwarfs that of net firm entry and exist. Clearly new firm formation is only one dimension of innovation and existing firms account for a large share of total research and development (R&D), in many industries such as pharmaceuticals. It seems US any industries will need spin off of existing operations and the acquisition of independent start-ups are now important dimensions of new process and product development. Thus, innovation is an important factor to influence US economic growth in the future.

However, it will have challenges to US different industries when the causal connections between entrepreneurship, innovation and growth arises. For example, US employment growth of large, middle and small sizes of entrepreneurships will be strongly positively correlated with US any new firms formation, but this doesn't necessarily imply that entrepreneurship causes growth . There may be one factor that causes US employment growth and US firms formation to co-vary, and it is hard to find instructions that affect US firm formation, but have no independent effect on US employment growth.

The interpretation of the correlation between US employment growth and US firm formation relates to old debates in US economic geography about whether US workers follow US firms, or US firms follow workers or there are mutually reinforcing feedbacks between US firms' and workers' locations decisions. So, some economists believed that, US any domestic geography economic growth will possible be caused by these factor , such as either US geography firms innovation factor or US geography high technological workers high level productivity factor or US geography firms innovation or US geography high technological workers high level productivity both combining factor. But the challenge will cause, it assumes that future many US entrepreneurships choose to innovate to manufacture whose products. Although, many high technological manufacturing workers employment chance will be raised, but it also cause many low technological manufacturing workers employment chance will be reduced as the same time. So, it is difficult to keep high technological and low technological both workers who have same fair employment chance as the same time in the future US society if US planned to achieve the future knowledgeable innovation economic society.

However, to achieve this innovation dream, the economists give opinions to US government which will need to encourage the foundations of US entrepreneurial policy and distinguish four broad actors: (a) individual agents who identify business opportunities and choose to exploit them. (b) new formed businesses which innovate using new knowledge and other resources, (C) the economy including all institutions that influence economic growth, and (d) US society is as the collection of all agents who are the ultimate beneficiaries of wealth creation. Within this organizing framework, US entrepreneurships will be easily shaped or changes the overall business climate to achieve knowledge economic society in the future.

The role of intangible assets influences the regional economic growth in US.

What are the four big factors of intangible assets? Some economists indicated they include knowledgeable capital, human capital, social capital and entrepreneurs capital. Nowadays, globalization and increased competition will cause new types of pressure to influence US economic growth. So, US companies need have flexibility, the ability to immediately adapt to market developments, and pro-activism in creating future markets. The relative importance of physical growth: However, soft production factors, that is those related to personal knowledge are becoming more important to influence US future economic growth, which regards to human capital and knowledge as driving factor of economic growth in industries developed countries, such as US.

All these soft production factors can be grouped in what is known intangible assets. These assets can be defined as non-material factors that contribute to enterprise performance in the production of products or the provision of services, or that are expected to generate future economic benefits to the entities or individuals that control their deployment (Akerlof & Kranton , 2000).

Whether it has a close relationship between these intangible assets and US regional economic growth? Some economists had researched to have more reliable quantitative statistical information to do report in above four big factors how to effect on US regional economical growth influence. Their report indicated that returns from human and social capital are taken as homogeneous for US all regions. They also indicated these two sets of questions in their report. The first one, dealing with knowledge accumulation, addresses, among others the following issues:

(a) How does innovation and knowledge accumulation occur within firms and hoe does it impact on economic performance?

(b) What is the role of universities in regional, national and global knowledge accumulation processes?

The second set of questions addresses the key knowledge diffuses over space and how this diffusion impacts on economic performances.

In particular:

(c) To what extent knowledge diffusion is conditioned by spatial proximity?

(d) What is the impact of knowledge accumulation and diffusion on economic performance?

They indicated to answer these questions, having reliable measures of how innovation process occurs, of the actors that take part in it and the mechanisms that are in place is important. They pointed out that the systemic nature of US regional intangible asset demands indicators that grasp two kinds of capabilities: network capabilities , i.e. connectivity, both intra-and-inter-regional and organizational capabilities, as well as their dynamic in US. These indicators have been applied to the study of linkages and relationships between US firms and between US firms and US universities, highlighting the mechanisms through which those actors contribute to the processes of knowledge accumulation, generation and diffusion.

They also found results: How does knowledge accumulation occur within US firms and how does it impact on US economic performance? They included collaborations with competitors are most commonly undertaken at early stages of the development projects, those with buyers and suppliers are more likely to result in the introduction of new products and processes: At the same, it has been highlighted that, regardless of the industry, the most innovative firms, i.e. those more able to absorb knowledge are more likely to participate in collaborative networks.

Thus, US future entrepreneurships need to offer knowledge accumulation opportunities for US regional firms and research institutions to benefit and contribute to global network of US expects to achieve rapid regional economic growth. US's entrepreneurships also needs to explore organizational innovation, those US firms whose structure enables learning by doing, using and interacting by relying, among other things on parallel development teams. Semi-autonomous work teams and reduced management layers are more likely to introduce new products to the future US domestic and foreign both markets.

What is the role of US universities in US regional, national and global knowledge accumulation processes? US universities can heavily influence US regional, national and global knowledge accumulation processes. For instance, when collaborating with multi-national enterprises, they may affect simultaneously the three level. Their report had highlighted interesting results on the local impact of universities: when it has been found that top ranked departments are significantly associated with partnerships involving spatially close industry partners, it has also emerge geographical proximity, is nt the main driver of collaboration choices. These are found to depend largely on US firms' networks and universities' specific characteristics. Among other things, the cultural traditional of academic institution has been shown to influence the ability to collaborate with industry and commerce research.

Networks characteristics have emerged as crucial in determining academic knowledge transfer: the better the access to international networks, the higher the patenting activity, have the knowledge transfer to US industry. This implies that the set of tools of knowledge based economic development should include not only research and development, but also clever ways of supporting academic research network.

To conclude, if US expected research development can assist economic growth. Then, US academics universities and companies both need to co-operate to carrying on experimenting any kinds of research and prepare to assist any US product innovation more easily. So, the intangible asset, not as scientists, technology, US universities‘ and firms’ research fund which is an conditional factor to influence US regional research and development economic growth in the future.

Social factor impacts US
technological development growth.

Has it relationship between social influence and economic environment in future US? For example, the social factors that are positively correlated with the economic growth (i.e. the expected years of schooling and the life expectancy) and respectively, the factors that are negatively correlated with US future economic growth (i.e. the US population or risk of poverty and the unemployment rate).

The improvement of the US future economic environment will be an objective of the macroeconomic policy on short, medium and also long term. The importance of social factors upon US future economic growth, considering that the future used macroeconomic indicator, GDP per capita, is not most proper measure for the future US nation welfare. Due to GDP per capital fails to take into consideration some specific sectors of the US social economy, such as the black market.

Until recently, some economists rely on culture is as a possible determinant of economic phenomena. However, in current years, better techniques and more date made it possible to identify systematic differences in people's preferences and beliefs and to relate them to various measures of cultural environment suggest an approach to introduce cultural-based explanation that can be tested and are able to substantially understand economic phenomena.

The increased importance of social factors relies on a basic concept. Some theory is measured to economic growth which has wrong assumption. For example, the fiscal and monetary policies focused on increasing the national income, which lead consequently to economic growth. The reason most of economic opinions have been argued because whose opinions are based on a wrong hypothesis, according to which the nation welfare is based only on the level of income.

Can social factors influence US future economic growth? Human development history, global life expectancy has been experiencing these stages: from the industrialization process, the technologic progress, the medical evolution, the scientific research, these stages were also related to internal causes, specific to some developed countries, e.g. US developed country. Thus, the differences are significant and are linked both to US life expectancy level and the GDP /capita. Such as US population is less than China population too much. Although, US land area is near to China area. It seems US will encounter life expectancy level need to prepare its technological development to raise economy growth of opportunity. For example, Africa and Asia are still facing major economic and social issues. The access to a health life and medical services are still long terms objectives for countries with low life expectancy.

According to Harrison & Huntington (2000), the analysis of social factors helps understanding the human behavior with respect to consumption, savings, investment system, expectations and attitudes towards the economic circumstances, which also have a major impact on the economic growth. The evolutions of economic and social environment are needed for US future development. In order to eliminate the gap of living standard, outside resources and support US needs have good social indicators study plan to concern econometric model to rise poor people living standard in future US society between rich and poor people who are living in US. However, I believe the social factors include demographic and culture, population's structure factors which are one important social indicator to influence the distribution of the US social public income.

Barro & Sala-i-Martin (1996) defined culture is as the sum of symbols, meanings, habits values, behaviors and social artifacts which characterize a distinctive and specific human population group. For decades, economists and social thinkers debated the influence of population change on economic growth. Bloom et. al (2001) defined three alternative hypotheses: that population growth restrict, promotes or is independent of economic growth . Each hypothesis was sustained with strong arguments, and all the arguments mostly focused on population size and growth. The debates revealed other important issues, such as the age structure of the population, the way in which the population is distributed across different age groups.

The economists indicated that people's economic behavior varies at different stages of life, changes in a country's age structure can have significant effects on its economic performance. So, such as US, if it had a high proportion of children are likely to devote a high proportion of resources to their care, which would tend to depress the pace of economic growth. By contrast, if US's population falls within the working wages, the added productivity of this group can produce an increase in the economic growth. This is how the combined effect of this large working age population and health, family, labor, financial and human capital policies can create cycles of wealth creation. On the other hand, if US a large proportion consists of the elderly , the effects can be similar to those of a very young population; a large share of resources is needed by a relatively less productive segment of the population , which likewise can inhibit economic growth in future US society.

Further Bloom et al. (2001) analyzed the three main mechanisms of population's structure for determining economic growth (labor supply, savings and human capital) and their dependence of policy environment. They indicated that a growing number of adults will only be productive of there is sufficient flexibility in the labor market to allow its expansion, and macroeconomic policies that permit and encourage investment, people will only save if who have access to adequate saving mechanisms and have confidence in domestic financial markets and the demographic transition creates conditions where people will tend to invest in their health and education, offering great economic benefits, special in the modern world's increasing sophisticated economies.

It seems future US labor market will need a growing number of adults to supply , a stable domestic financial market to encourage US people have more confidence to save money in banks, a stable health and education sector industry can encourage US people have confidence to invest to do this kind of health and education service industry. Instead of population of working age factor will influence US economic growth. However, US people living culture will also influence US economic growth. For example, if it has a trend that there are many US young people like often go to shopping in relax time, because who feel shopping is their young group entertainment in the year. Then, the year US GDP growth will be possible raised, due to there are many US young people prefer to spend for shopping expenditure suddenly in the year. In conclude, in social environment of cultural and population of working age size will influence US future economic growth.

Online tourism industry influences US
future online industry economy growth.

Today, US online tourism sale industry has always been one of America's great home growth industries. Today, more than 8 million Americans are employed in travel and tourism. For example, US domestic South Carolina, hospitality and tourism has been as the largest local industry, providing tens of thousands of families directly or indirectly with jobs. Predicting global tourism consumption needs work be increased. US will experience in a changing economy, tourism industry will provide job security to many Americans as well as the service -oriented nature of many travel positions and these jobs are difficult, if it is impossible to outsource.

Many people accept to buy air tickets from online sale channel. Will future America government play an important role to influence future US online air ticket tourism sale method to raise online travel consumption behavior? However, some have questioned in recent years whether US policies have harmed the ability of the tourism industry to expand. One of the most frequently discussed concerns involves American visa policy, when European Union allows tourists from 26 European countries and almost all of the America and Australia to visit without a visa. America policies require some travelers from friendly countries, such as Brazil, to travel thousands of miles from home to attend the in-person interview needed to secure a tourist visa. Although, the US White House has taken steps to reduce rise wait times in recent years, in some parts of the world, like Vietnam and Turkey, the US tourist visa application process can be a multi-month process (US Travel Association , 2014).

Will the visa waiver program impact driving increases in US impact driving increases in US tourist volumes? Some economist analysis found that when a typical country joins the visa waiver program, it sees a notable increase in the number of countries who chose to visit the US in the immediate years. That follow over the course of its first five years in the US visa wavier program, the number of tourists arriving from a participating country rises by 16.4%. Also their analysis indicated to expand the visa waiver program to some countries would result in $7.66 billion in additional tourist spending within a five year period. It would also create at least 50,000 American tourism jobs within 5 years. Purchases of products and services by visitors contribute foreign significantly to job creation and economic growth in US international travelers to US purchased more than $180 billion amount . In addition, tourism generates a trade surplus as foreign visitors spend more in the US. So, US enjoyed a $57 billion trade maintained a trade surplus in tourism every year. Some economists indicated that India is likely to contribute to the growth in foreign travel to the US . If India government can ensure which have enough numbers of foreign travelers are encouraged to choose to travel to US. In the future, India will be one important tourism cooperation partner to US. So, US ought achieve online travel sale strategy to promote to Indian to know what online travel advantages can give to them in the future, e.g. reducing time to buy air ticket, convenience, cheap e-ticket price, avoiding to lose air seat supply, more airlines choice providing. These are online e-ticket sale attractive characteristics to India online e-ticket sale market. Thus, US ought concentrate on promoting this kind of online travel sale service to India in order to raise US many airline e-tickets sale numbers.

The effects of population growth influence US economy growth.

Nowadays, China and India population growth rate are rapidly. Population growth can have several disadvantage effects on the economic expansion and performance to China and India. Does it influence cultural and sociological links between the following per capita income, rates, technological advancement , education aspects? China with 1.32 billion people and India with 1.1 billion people. Though, they each enjoy a large labor force advantage, several key economic factors have contributed to how the Chinese and Indian populations have grown and what differing effects that growth has had on their developing economies. Although, there two countries have high population growth rates, but the poor people has low standard of living.

So, when these two countries population sizes are large, but the standard of living will be low, and population will be reduced by either the preventive check (international reduction of fertility) or by the positive check (malnutrition, disease, and famine). It seems china and India governments think overpopulation challenge will cause social crime increasing, education competition, raising the job applicants number will increase, but the jobs supply number won't increase faster than job applicants number, and it will cause unemployment etc. social challenge in the future long term. So, why China and India governments will choose to control birth rates every year. However, in these developing countries, children are often as an economic asset, a tool to help increase agricultural production. Demand or need for large number of children is driven down by improvements in living standards and child survival and by the modern of economies. The movement away from agricultural production and into more modern industries in the manufacturing and services sectors need the necessity to have numbers of children to prepare to supply to China and India future manufacturing and services job markets.

However, US is a developed countries. The country main GDP growth source is high technological industry. Hence future children birth rate is not necessary to prepare to supply for manufacturing or service industries needs mainly. It seems US population birth rate is stable. So, overpopulation challenge won't be caused easily. Can population growth influence US economic growth? US is different to China and India. US's population growth rate is less. Hence, if US population growth rate can rise. It won't have overpopulation challenge. US needs have many young people number who can have high technological knowledge to prepare to have enough supply to do the future high technological different kinds of positions to build social economic growth in the future. Hence, US ought encourage young people choose to marry and to born next generation children to supply future different kinds of high technological jobs of needs to raise technological productivity and they (high technological manufacturing workers) will be the main production of factor to influence future US economic growth in the high technological manufacturing sector.

Thus, I believe the effect future US of population growth will influence US future economic growth as the same time. Also, it is the right time, US government needs to assist universities have enough afford resources, e.g. university fund, technological educational courses, lectures etc. to provide future US technological education students' needs in the future. So, it brings this question: What can determine US economic growth?

In global macro-economic influence, since 1973 year, per captia income growth in the US and other advanced countries has slowed to 2.2 % a year or almost half the 3.9% annual rate of the preceding quarter century . If the US had maintained the level of growth experiences in the 1950 year and 1960 year, real per capita income today would be about 71% greater than it actually is. In contrast, it has been estimated that eliminating the variability in US consumption since world war II would be equivalent to boosting current real consumption by only about 4.8%. If the choice is between long-term growth policies and further short-term stabilization policies long term clearly have the potential for higher benefits. Hence, what factors determine US economic growth (David M & Roy J, 1993).

According to traditional growth theory, the main determinants of long run economic growth are not influenced by economic incentives. Perhaps the reason why economists have neglected long time, to profession relied on a theory that offered litter scope for policy to influence important sources of growth.

High technology development can
reducing income inequality growth.

Some economists found one key is that education and anti-discrimination policies well designed labor market and large and/ or progressive tax and transfer system can all reduce income inequality . In many OECD developed western countries, income inequality has increased in past decades. In some countries, top earners have captured a large share of overall income gains, when for other income has risen only a little. Some see poverty as the relevant concern with the type of growth enhancing policy reforms advocated for each OECD developed countries and economic growth might have positive or negative side effects on income inequality.

OECD (2011), it first highlights differences in some income inequality across the OECD and the factors driving them, such as cross-country differences in wage and non-wage income inequality as well as in hour worked and inactivity. OECD developed western countries can be divided into five groups to their pattern of inequality. For example, in five English-speaking countries (Australia, Canada, Ireland, New Zealand, the United Kingdom) and the Netherlands wages are rather dispersed and the share of part-time employment is high, driving inequality in labor earnings above the OECD average means-tested public cash transfer and progressive tax.

It seems income inequality will influence the developed countries, such as UK, Canada, Australia, New Zealand , US etc. economic growth. Although, technology change and globalization have played a role to influence the distribution of labor income. Some economists believe that any countries' policies will also influence income inequality. These policies factors include: technological education policies can increase different technology graduation rates from upper secondary and tertiary education and that also promote equal access to a well-designed different sector technology labor market policy can reduce inequality.

A relatively high minimum wage narrows the distribution if labor income, but if set too high, it may reduce employment of inequality reducing effect. It tends to reduce labor earning inequality by ensuring a more equal distribution of earnings. Job protection reforms that make permanent and temporary contracts more even in their provisions low income wage dispersion of earnings is rather mixed, removing product market regulations can reduce labor income inequality by boosting employment, policies the faster the immigrants and fight all forms of discrimination reduce inequality, progressive tax policy play a key role in lowering overall income across the OECD developed countries.

However, the redistributive fair income level between low level income labors and middle level income labors and high level income labors impact of developed countries, e.g. consumption taxes and real estate taxes tend to be regressive tax policy. Hence, it seems that reducing income inequality can cause the more fair income distribution between the high income level and the middle income level and the low income level labors. Then, it will let the developing countries or developed countries , such as US citizens feel that who can get real social welfare fairly, due to high technology development can boost economic growth in US society.

Long term cheap medical cost trend factor influences US economic growth.

Some economists predict to make long term forecasts to reduce medical cost trends how will influence US economy. Also, they indicate short run cheap medical cost forecasts for first 1 to 5 years to reflect the particulars of specific groups, benefit packages, regional markets or cheap medical cost network providers and use their local cheap medical choice information and actuarial skills to improve accuracy and reasonability. They also oversight group review on the consistency of baseline assumptions with factor influencing future US patients‘ medical choice taste, medical technology trend component and historical annual percentage increase in medical costs, premiums, income and excess cost growth are found to influence future US patients consumption behavior. However, their research predicting expensive medical cost is the main factor to influence to reduce US patient numbers among of ward room sleeping environment, doctor loyalty, hospital entertainment and service facilities provision hospital cheap car parking charge service, cheap or free telephone call provision, hospital meal quality provision etc. different factors .

They suggest that how to reduce medical cost. They indicate many other factors, e.g. patient individual illness aging, physician supply, medical insurance plan benefit will affect the cost of a particular patient, group, organization or plan in a specific locality or over the short run. Also they indicated that additional analysis is still needed to determine if reliable estimates of separate medical insurance premiums, medical care payments, pharmaceuticals or other cost categories are possible to influence medical cost to be increase.

Hence, for long run US cheap medical cost will encourage or attract many patients prefer to pay medical cost to see doctors in US any hospitals or medical centers. Then, US hospitals or medical centers income will increase and US GDP income will also increase in medical service sectors in the future. It seems that long run cheap medical cost will influence US economic growth.

Long term cheap medical cost trend factor
influences US economic growth.

The next generation of US talent management practices and solutions will largely be driven by US economic evolution, demographic changes and technology advancements. These factors are dramatically influencing the way US people work, the way US companies are organized and the way talent is managed.

Some economists explained the key economic factors driving changes in talent management include: the knowledge economy is value to companies, talent is now a required strategic asset. Key changes in the future include a line between inside and outside talent that will result expansion of the talent management, globalization, such as European expansion is well-known top expansion prospect for global companies now include China, Russia and Eastern Europe and America and the rest of Asia. So , US government needs to know globalization can give what opportunities to US future .

Although, knowledge management or knowledge economy will bring advantages to US. However, it also bring disadvantages to this aspect, such as skill gap and structural unemployment will also influence US organizations structural unemployment and skill gap challenges issues will be caused between low skillful and high skillful workers, generational geographies changing will occur in US.

Although, baby boomer retirement has been top of mind for many years in the US, even more significant demographic changes are happening outside the US, where population growth rates and aging population will influence as local economies. Thus, the ability for organizations success will need global talent or effectively more talent from areas of abundance to scarcity is becoming a strategic issue to any US companies development , increased health and longevity mean that seniors are working longer enabling US organizations to keep experienced team members into their retirement years. But it also raise US workforce planning and generational challenges, workplace and diversity is increasing, a more diverse pool of talent affords new opportunity . Such as hiring workers raise productivity of needs, digitization of candidate or employee profiles can meet US business employment needs because employee talent data has been digitized and integrated into comprehensive talent profiles.

Techniques , such as attribute matching and recommendation technologies can be applied in talent management can be applied in talent management to find and match the most right candidate to do any positions in short time efficiently, with the increasing need of telecommunication in the US market penetration of smartphones and tablet devices a significant portion of the world's human potential will have access to rich web and application experiences from anywhere. Thus enables US organizations to source and collaborate on knowledge work with any part of the world into a global talent pool.

How can intellectual assets impact to US to change to knowledge economy? The information age moved the basis of economic value from products to intellectual assets, information and the talent that develops them. It is now widely acknowledged that intangible assets, with largely consist of know-how, unique intelligent property, and patent right, drive move than 80% of the valuations of publishing US publicly traded companies.

In future US knowledge economy, US leading edge organizations will have efforts to use into the talent and intellectual capital of not only employees, but also clients, partners and the public at large in an effort to create an extended electronic community. For example, client support portals where clients can answer each other's questions, which will reduce costs by expensive client support calls for US companies in future US knowledge economy environment.

Also, knowledge economy will assist US companies to be partner and customer innovation efforts easily. Such as P&G daily product company's ability to source more than 50% its new product ideas to external innovation, driving industry-leading standards for new product launch rates. In future knowledge economy environment, it is clear that US organizations will be increasingly deriving value from talent that is outside the company.

In future US knowledge economy environment, the link between employees within the US organization and those outside, it is also driving talent management changes, particularly around sourcing, strategic workforce planning and employees engagement . Given these changes, executives in leading US companies will be increasingly focused on talent management issues, recognizing that talent, wherever knowledge economy will give knowledge management method to any US organizations to build more competitive advantages in the future. However, knowledge economy will also bring structural unemployment to US society. Unfortunately, today's global talent market is largely inefficient and characterized by a high degree if competition relative to the redistribution of talent.

One result of this inefficiency is long period of skill gaps structured unemployment is a particularly different scenario in which section skills are no longer required, not just within a US particular company, but within US an entire sector. Structural unemployment may be the future result of cyclical boom and bust cycles, offshoring of partial. In conclusion, whatever the causes of future knowledge economy to US , the result is an uneven distribution of talent relative to the available jobs in US. In part of structured unemployment in US will be caused from knowledge economy influence. However, knowledge economy will also influence US future businesses speed increases, bandwidth increases and growth in mobile internet access during US encounters the knowledge economy influence in society.

The impact of educational quality factor influences US economic growth.

Schools can train skills of an individual and human capital for future businessmen need. It is not the only factor. Schools nonetheless have a special place, not only because education and skill creation are among that prime explicit objectives, but also because which the factor(s) most directly are affected by public policies. So, US is well established that the distribution personal incomes in US society is strongly related to the amount of education US people have had. Thus, any noticeable effects of the current quality of US schooling and the distribution of US people skills and income will become apparent some years in the future, those US students now in US school become a significant part of the future US labor force. Then, the question arises as: To whether are these skills correlated with US student's subsequent performance in the future US labor market and will the US economy's ability to grow ?

The quality of human resources are measured by scores is directly related to individual earnings , productivity and economic growth. A range of research results from the United States shows that earnings advantages , due to higher achievement on standardized tests are quite substantial . These studies typically find that measured achievement has a clear impact on earnings, after allowing for a differences in the quantity of schooling, age or work experience, and for other factors that might influencing US people earnings. In other words, for those leaving school at a given grad, higher quality school outcomes (represented by test scores) are closely related to subsequent earnings differences. So, it seems quality of education and individual income has direct relationship to influence US economic growth.

According, educational program that deliver those skills will bring higher individual economic benefits. Obviously, students who don't better in schools as evidenced by either examination grades or scores on standardized achievement tests tend to go further in school or university. By the same way, the net costs of improvement in school quality if reduction in rates of student repetition study. Then, it brings this question: Will education quality bring future economic return to US?

Early some economists' research found that personal and behavioral traits, such as perseverance and leadership qualities had a significant influence upon labor market success, including earnings. To answer above question, we need to know if US schools which generally have good student performance, it will influence socio-economic advantage improved performance, changes in school climate, teacher morale and commitment, school autonomy , teacher-pupil relations and disciplinary had some compensatory influence towards greater equity. In American, pupil socio-economic background and classroom climate appeared to be the most important predictors of achievement. So, it seems human development will influence US economic growth in the future. So, US development organizations seem better will then non-specialized educational organizations to design and deliver effective combinations of livelihoods and literacy.

Whether people who attain literacy actually make much use of it is subject to debate. On balance, however, literacy seems more used where economic development is better established. This supports the argument that same degree of economic and political improvement is necessary to sustain literacy. Then, US people will use their literacy skills where conditions make it useful or desirable for them to do so. However, schooling is a social process, and improvements in resources, technology and quality of student and teaching inputs should in principle to able to be enhance its overall quality. In a good number of countries, large increases in average real expenditure per student and other measures of school resources in primary and secondary schools over the last four or five decades have not remotely been matched by a comparable increase in average test scores.

What can be influenced from quality of education to US? Some economists suggest the earning can provide an important incentive mechanism, which can influence both the quality and motivation of teachers. If teachers‘ real average earnings had kept the same level with other professional groups over the period, the productivity impact of their earnings growth would literacy have been small. The first view point, in fact, however, in many countries teachers' earnings have increased considerably less sharply than those other groups. Teachers may feel worse off, because of their decline in status relative to other professional groups. This circumstance could well explain part perhaps an important part of the apparent lack of impact on learning outcomes of increase in real per student spending over time. One to teachers‘ general salary is low to compare other professional in the global society.

Then, it brings this question. Can quantity of education influence US economic growth? In this modern economic approach to investigating the determinants of educational outcomes has borrowed well established technique from other economic applications. The idea that there is a determinate relationship between (teacher number) inputs to a production process (teaching process) and the outputs(student examination results and learning performance) that subsequently bring has being been important in micro-economic analysis. If US schooling teaching possibilities are governed by certain teachers' quality between factors of production, e.g. teachers; teaching performance , teachers' supply numbers to every schools, teachers' educational qualification. So, the product function describes the maximum feasible output (student examination results and learning combinations) obtained from alternative combinations of these inputs (teachers' quality teaching performance, supply numbers, education qualification).

Production factors are powerful analytic tools, which have been applied to the analysis of most forms of economic production. Since the mid-1960 year, they have also been widely used in the economic analysis of education. Thus, in schooling educational organizations, teacher salaries will influence teachers' performance (production of factor) to be good or bad. If teachers' educational performance is good, then production of output (student learning abilities and students examination results) will be good usually. Otherwise, if their educational performance is bad, then production of output (student learning abilities and students examination results) will be bad usually .

In US society, if many poor learning performance students have been produced. Then, there will not many talent young people who can contribute to serve US society in different professional aspects. It means that they can not contribute US economic growth easily. Thus, it seems that US quality of education can impact US economic growth for long term and US universities, secondary schools and US government ought need to find methods how to raise quality of education for next generation of labor supply market.

Bio-medical industry factor boosts US economy.

In the future, the perspective of bio-medical industry will be an alternative growth scenario. So, US government will need to modest improvements in key policy areas to adopt bio-medical industry needs, e.g. more favorable coverage and payment policies for medical innovation, improvements in regulatory policy to create efficiencies in research and development process and improvements in policy to incentive R & D (research and development) create efficiencies in the R & D process.

As an industry rooted in science and advanced manufacturing, the innovation bio-medical industry is uniquely position to help maintain US leadership in new technologies and scientific to continue to create high quality, high wage R&D and manufacturing jobs and enhance America's global competitiveness in the future. Today, the US bio-medical industry supports a total of 3.4 million jobs across the US economy , including over 810,000 direct jobs: contributes $189 billion in economic output and is responsible for about one in five dollars spent on domestic R&D by US businesses. Hence, innovation bio-medical industry will be one high value knowledge-based industries as a driver of US economic growth.

The reasons include: The first reason is such as , the US history, it has been the world leader in bio-medical research and development of new medicine over the past 30 years. It has been one leader of world class life science ecosystem and innovation among different countries in medicine development history. So, bio-medical industry will be a high technological innovation related activity of industry leader, as is measured by R&D investment, e.g. generation venture capital and share a total R&D employment in manufacturing industry. Next reason is that US bio-medical industry employs a total of 813, 523 worker amount. These workers lead a wide range of occupations that offer high wage, high quality employment. The total economic impacts of the industry, it is estimated (via the generally accepted methodology of input/output analysis) that supports nearly 3.4 million total jobs and generates nearly $ 789 billion in US economic output (National Science Board, 2014). Finally, reason is that US people concern life health issue, In addition to improving individual health and lengthening life ages, medical advances have contributed to substantial societal health gains, such as reducing disability and improving productivity.

According to two Universities of Chicago economists indicated the estimated economic gains from declining morality alone in the US from 1970 year to 2000 year had a value to US society of more than 3 trillion a year. Hence, US medical health need of patient numbers will increase, due to who have large needs for life health. Hence, quality of education can influence bio-medical industry development to impact US economic growth in the future.

CHAPTER FOUR

Future technology development to human occupation

Economists explain that welfare economic theory, Kaldor-Hicks efficiency theory and Pareto improvement theory which have close relationship. Welfare economic theory means general social equilibrium. They indicate the demand of commodities must same to the supply of commodities when the commodity market stays in the stable level. So , when production is increasing, the manufacturers also need to keep their products supply and demand keep to the stable level and consumers' demand of their products must also need to increase as well as bank interest rate must be fallen in order to raise consumption desires. Also, the low bank interest can also excite investors' investment desires.

Such as to explain why the developing countries cities technological competitive investment factor can impact US and UK economic growth. Welfare economic theory means general social equilibrium. The economists indicate the demand of commodities must same to the supply of commodities when the commodity market stays in the stable level. Due to the future developing countries investors will invest much on developing new technological products research. So, these countries' economy growth will be possible faster than US and UK both developed countries because it is possible that the technological product investor number will decrease. So, US and UK government will need to attract and encourage many foreign developing countries' investor prefer to invest their technological products research to assist US and UK both developed countries to develop its economy. As Welfare economic theory means general social equilibrium. If US and UK government can give benefits to these developing countries' technological product investors, then these developing countries' technological product investors will prefer to invest to developed country US urban cities, such as small cities or farming location to build factories to manufacture their new technological products. Because , London, New York, Washington etc. large cities' technological investment and location where land supply is limited and offices and factories rents are also expensive. Otherwise, UK and US, farming location or small cities location where have much land supply to build large offices and facilities and the rent is much cheaper than US and UK main cities, such as London, New York, Washington etc. Even, if UK and US government and banks can lend low interest rate of bank loans or government loans to encourage foreign investors to do technological product research in its country. So, UK and US bank low bank interest rate can encourage US consumers prefer to buy any future new technological products, due to low bank interest rate can not attract who choose to save more money in US and UK any banks as well as US and UK bank low bank interest rate can also encourage foreign developing countries' high technological product investors to choose to invest in UK and UK , due to who can pay low bank or government loan interest to compare themselves countries or other countries.

As Kaldor-Hicks efficiency theory indicates the policy is suitable to be implemented if the policy beneficiaries' welfares can compensate to the policy benefactress. So, US and UK bank and government low loan interest rate policy is valuable to implement if benefactress, such as US and UK government which can raise economic growth as well as the US benefactress, such as the high technological product businessmen can sell many different kinds of high technological products from foreign developing countries' investors' assistance in US and UK both developed countries.

As Pareto improvement theory indicates that it can not carry on continue, due to the country has shortage of natural resources after the country improved its welfares to reach the maximum standard. The result will be caused, such as the country will reduce this group of citizen's welfares, if the country decides to continue to improve another group of citizen's welfares. So, future US and UK will have shortage of natural resources to supply to technological product manufacturers. So, the number of technological product production will be fallen , due to the shortage of natural resources are supplied to US and UK high technological product manufacturers. The only solvable choice is encouraging foreign high technological product investors to attract them to supply themselves high technological products any resources to manufacture high technological products and provide to US and UK high technological product sellers to help them to sell in US and UK domestic market or export to overseas market to earn income. Then, US and UK will earn raise GDP income from high technological product sale industry in the future.

Natural rate of unemployment

Economists believe why natural rate of unemployment will occur in any countries. They indicate that during economy condition keeps at the balance situation (condition), anyway any country government adopts any policy. However, the natural rate of unemployment won't reach to zero level as well as there are some people will still unemployed or it is possible that some people will be alternative employment between any time. Hence, it means that although US and UK is an developed country. It can not guarantee there are not any people unemployed , so the natural rate of unemployment will not reach to zero level.

So, it seems that US and UK current economy condition had kept at the balance situation (condition). Thus, even future science, technology , engineering and mathematics workers will drive US innovation and competitiveness by new ideas, new companies and new industries in US. However, US and UK employers frequently concern the supply and availability of this kind of workers. Over the past 10 years, growth in this kind of jobs was three times as fast as growth to general jobs in US. In the future the Natural rate of unemployment of US and UK science, technology , engineering and mathematics worker number must not reach zero level.

Anyway, in the future, the perspective of bio-medical industry will be an alternative growth scenario. So, US and UK government will need to modest improvements in key policy areas to adopt bio-medical industry needs, e.g. more favorable coverage and payment policies for medical innovation, improvements in regulatory policy to create efficiencies in research and development process and improvements in policy to incentive R & D (research and development) create efficiencies in the R & D process.

As an industry rooted in science and advanced manufacturing, the innovation bio-medical industry is uniquely position to help maintain US leadership in new technologies and scientific to continue to create high quality, high wage R&D and manufacturing jobs and enhance America's global competitiveness in the future. In the future the Natural rate of unemployment of US and UK of bio-medical industry worker number must not reach zero level.

Thus, if future US had enough job number which could supply to these high technological and bio medical industries of labors to do. However, the natural rate of unemployment must not reach at zero level and it does not represent its economy is poor because one developed country , such as US had been experiencing economy condition keeps at the balance situation(condition), so it's natural rate of unemployment must not reach zero level . Otherwise, the developing countries' economy condition does not keep at the balance situation (condition), so their natural rate of unemployment have more possible to reach to close zero level.

Industrial production and pollution economy

Economists indicate that it has close relationship between industrial production and pollution. When traditional products are manufactured, the air and water pollution will be caused in the economic activity. So, human needs to protect environment to keep health, the demand of investment of money reduces pollution and labor number will increase to achieve to reduce air and water pollution in natural environment. Thus, product manufacturers need to analyze which economic stage(s) will encounter shortage and find methods to solve challenge.

This Industrial production and pollution economy theory can explain why US and UK needs to plan achieve digital economy to shift future America rural economy. Some economists indicate that there are five trends reshape to impact rural America' and England future economy. They include that digital economy will shift future America rural economy. US and UK quality of life will change a lot, the US and UK rural economy will stay uneven, US commodities will compete in global markets and will give less benefit to US and UK rural economy and US and UK new products will revolutionize US and UK agriculture economy.

US and UK needs to plan digital economy to future agricultural production of the reasons include: The first aspect impacts to US agriculture economy, the US future rural economy stays uneven. Growth will concentrate in 4 out of 10 rural places and they have scenery, a retail hub, or one next to a city in US and UK. The impact on rural America includes some rural places will try to manage growth , but many places on a quest for new economic engines. Thus, it will bring these questions to US and UK rural economy impact, such as : Who will be US and UK businessmen clients? The struggling farmers? The struggling farm-dependent country? The booming mountain area? The rural area transforming into city? The second aspect impacts to US and UK agriculture economy, due to US and UK commodities will compete in global markets if it will bring a smaller benefit in US and UK rural economy. Then, US and UK farm scale will cut costs and competition fewer US and UK farms and places will depend on farm income. The third aspect impact to US and UK agriculture economy, new products will revolutionize agriculture. Major, shift from commodities to products, spurred by biotech, means two agriculture in the future and two rural America . Hence, US and UK future agriculture determination the rural economy will be declined. A future new US and UK agriculture supply chain integrator will be caused from the traditional farming supply chain procedure the change to outsourced contractor farming supply chain procedure, such as: In beginning, from farmer will outsource supply chain contract to processor, then contract to distributor and contract to food retailer final step. The fourth aspect impacts to US and UK agriculture economy, what will be two agricultures impact to the US and UK future economy? The first US and UK agriculture impact will be US and UK commodity agriculture. It focuses on production capabilities, farming foods production will be thin margins maintained with technology and big sale. The second US and UK agriculture impact will be product agriculture. It focuses on consumer needs, farming foods production margins will be protected by capturing value and building business relationship. In the fifth aspect impacts to US and UK agricultural economy, it is digital economy impacts US and UK rural agriculture. Future diversity economic base can encourage US and UK farming product agriculture to enter rural digital agricultural service consumption sector to change US and UK agricultural consumer individual shopping habit.

Thus, future US and UK agricultural production needs to concern how to reduce air and water pollution to influence natural environment clean quality as well as it needs to find what economic stage of agricultural production will be innovated in the future.

Reference

Akerlof, G.A. & Kranton, R.E. (2000): Economics & Identity, Quarterly Journal Of Economics, 105(3) , 715-753.

Barro, R., Sala-i-Martin, X., (1996). The classical approach to convergence analysis. The economic journal, vol. 106, no. 437.

Bernard, A., Redding, S & Schott, P. (2006). " Multi- product firms and product switching, " NBER working paper, 12293.

Bloom, D., Canning, D., Sevilla, J., 2001. Economic growth and demographic transition. National Bureau Of Economic Research.

Brynjolfsson and Mc Afee, (2011). Race against the machine: how the digital revolution is accelerating innovation and driving productivity , US.

Bresnahan, T.T. (1999). computerisation and wage dispersion: an analytical reinterpretation. The economic journal, vol. 109, no. 456 pp. 390-415.

David. M Gould & Roy. J. Ruffin. What determines long run economy growth? Economic review, second quarter, 1993.

ESA (2010 & 2018 year) calculations using current population survey public use micro date and estimates from the employment projections program of those Bureau of labor statistics.

Fernald, John (2015). " Productivity snd potential output before during, and after the great recession" In Jonathan A Parker & Michael

Hakkio, Craig S. 1992. " Is purchasing power parity a useful guide to the dollar?" Federal reserve bank of Kansas city, Economic review, third quarter, pp. 37-51.

Harrison, Huntington, S . (2000). Culture matters: how values shape human progess. Basic books.

John, G. 2016 " Reassessing longer run US growth. How low? " Federal reserve bank of San Francisco working paper 2016-18. http://www.frbsf.org/economic-research/publications/working-papers/wp2016-18/pdf.

Martin, J.P. & S. Scarpetta. " Setting it right. employment protection, labor reallocation and productivity." De Economist , 60: 2 (2012): 89-116. Online at: http://ideas. repec.org/p/iza/izapps/ pp.27 html (3).

MGI (2013). Disruptive technologies: Advances that will transform life business and the global economy. Tech. Rep: Mckinsey Global Institute.

National Science Board, 2014 , Science and Engineering Indicators, USP TO Patent applications and grants by industry.

OECD (2011), Divided We Stand: Why Inequality Keeps Rising, OECD publishing.

US Travel Association " US Travel Employment Reaches An Time high" (press release) , Nov. 7, 2014 Accessed Dec. 8 , 2014. available here : http://www.ustravel.org/news/press-releases/travel-industry-employment-reaches-all-time-high.

US Bureau of the census, 1993. Statistical abstract of the United States, 113 the ed. Washington.

CHAPTER FIVE

Can technology influence our living of standard

Technology innovation will change businss models to impact on the employment. Many of the major drivers affecting global industries are expected to have significant impact on jobs, ranging from significant job creation to job replacement and from heightened labor productivity to widening skills gaps. This global technology innovation brings this question: How will technology change impact on human job skills?

Economists predict that there are many industries and countries the most in-demand occupations or specialists did not exist ten or even five years ago and the pace of change is not to accelerate. They also predict that by one popular estimate 65% of children entering primary school today will altimately end up working in completely new job types that don't yet exist.

In such a rapidly employment needs change, the ability to anticipate and prepare for future skills requirements, job content and the aggregate effect on employment is increasingly critical for businesses, governments and individuals in order to fully seize the opportunities presented by these trends.

The World Economic Forum's Future Of Jobs Report seeks to understand the current and future impact key disruptions on employment levels, skill sets and recruitment patterns in different industries and countries. It indicates that human is today at the beginning of a fourth industrial revolution. Technological development will be in these fields , such as artificial intelligence and machine learning, robotics, nanotechnology, 3D printing and genetics and biotechnology. For building technology example, smart systems, homes, factories, farms, entire cities will help tackle problems ranging from supply chain management to climate change. Concurrent to this technological innovation is a set of broader socio-economic, geopolitical and demographic developments.

II. What will be future human employment trends?
The global workforce is expected to experience significant change between job families and functions. Across the countries are covered by the report, current trends could lead to a net employment impact of more than 5.1 million jobs lost to disruptive labor market changes over the period 2015 to 2020 year, with a total loss of 7.1 million jobs, two thirds of which are concentrated in routine white collar office.
Future of jobs survey, World Economic Forum indicated that the drivers of change , industries overall, share of respondents rating driver as top trend percentage. Such as demographic and socio-economic: changing nature of work, flexible work 44%, middle class in emerging markets 23%, climate change, nature resources 23%, geopolitical volatility 21% , consumer ethics, privacy issues 16%, longevity , ageing societies 14%, young demographics in emerging markets 13%, women's economic power, aspirations 12%, rapid urbanization 8%. Technological industry includes mobile internet, cloud technology 34%, processive power, big data 26%, new energy supplies advance technologies 22%, internet of things 14%, sharing economy, croud sourcing 12%, robotics, autonomous transport 9%, artificial intelligence 7%, advance manufacturing , 3 D printing 6%, advanced materials, biotechnology 6%.
Some economists also predict how the timeframe to impact industries, business models: The impact is felt already included: rising geopolitical volatility, popular mobile internet and cloud technology, advances in computing power and big data,

crowdsouring, the sharing economy and peer-to-peer platforms, rising of the middle class in emerging markets, young demographics in emerging markets, rapiding urbanization, changing working environments and flexible working arrangements, e.g. online working style, climate change, natural resource constraints and the transition to a greener economy. Then, from 2015 to 2017 year period, our life will be impacted by these influences. Such as new energy supplies and technology development, internet , advanced manufacturing and 3D printing, new consumer concerns about ethical and privacy issues, longevity and ageing societies, women's rising aspirations and economic power.

Further, in the future period from 2018 to 2020 year. Our life will be impacted by these influences. Such as advanced robotics and autonomous transport, artificial intelligence and machine learning and advanced materials , biotechnology and genomics will be mature developed. Thus, these technological changes will bring these impacts of this human job question, such as: How to change in ease of recruitment?

III. How to change in ease of recuitment?

Given the overall, disruption industries are experiencing, it isn't surprising that with current trends, competition for talent in-demand job families, such as computer and mathematical and architecture and engineering and other strategic and specialist roles will be fierce, and finding efficient ways of securing a talent technological development for every industry. Thus a tangible impact to further adequacy of employees' existing skill sets can already felt in a wide range of jobs and industries today. However, if skills demand is evolving rapidly at an aggregate industry level, the degree of changing skills requirement. Within individual job families and occupations is even more pronounced. Even, those jobs that are less directly affected by technological change and how a largely stable employment outlook.

How can technological development influence occupations or jobs future changes? For example, the mobility industries expect employment growth is accompanies by situation, where nearly 40% of the skills are required by key jobs in the industry are not yet part of the core skill set of these functions today.

As the same time, workers in lower skilled roles, particularly in the office and administrative and manufacturing and production job families, many find themselves caught up in a cycle where low skills stability means who could face redundancy without significant reskilling and upskilling even when disruptive change many erode employers' incentives and the business case for investing in such reskilling.

Future of jobs survey World Economic Forum also indicated skills stability from 2015 to 2050 year industries overall unstable and stable percentage such as below:

Media, entertainment and information industry skills stability 65% and skills unstability 35%,

Consumer industry skills unstability 30% and skills stability 71%,

Healthcase industry skills unstability 29% and skills stability 71%,

Energy industry skills unstability 30% and skills stability 70%,

Professional service industry skills unstability 33% and skills stability 67%,

Information and communication technology industry skills unstability 35% and skills stability 65%,

Mobility industry skills unstability 39% and skills stability 61%,

Basic and infrastructure industry skills unstability 42% and skills stability 58%,

Financial services and investors industry skills unstability 43% and skills stability 57%.

Thus, economists predicted the future skills serious shortage industries included basic and infrastructure industry unstable skills was 42% as well as financial services and investors industry unstable skills was 43%. Both were the highest percentage of the unstable skills among of them. It implied that university students could choose these both kinds of subjects to learn because these kinds of both industries will be shortage of graduated student number of supply. Otherwise, the other industries will have higher stable skills supply to employment market. So, it brings this question: How can stable and unstable skills influence to different industries and global employment market and living quality? For example: efforts to place unemployed youth in apprenticeships in certain job categories through targeted skills training may be self-defeating of kills requirements in that job category are likely to be different in just a few year's time. Indeed is some causes such efforts may be more successful if businesses models were on future expectations. Thus, businesses will need to put talent development and future workforce strategy front and centre to their growth.

It is therefore critical that long term changes to basic and lifelong education systems are needed to complemented with specific, urgent and forced reskilling efforts in each industry. This entails several major changes in how business views and managers talent both immediately and in the longer term. Thus, if university students can choose to study these any one subject among of them. Their living standard will be raised in possible, due to it will have many these kinds of job skills requirement in the future. Finally, it is also possible that unemployed young students will find jobs more easily and it will reduce young people unemployment number and it will raise life standard of future young people if nowadays any country universities began to encourage young students to choose to study any one of these above subjects to prepare to study . Thus, these industries' future unstable skills shortage will be decreased to influence any businesses development to be better in the future.

Chapter Five

Can technology influence UK and US life of standard to be fallen or grown

Some economists predicted long term global economic growth will be remain till to 2050 year at least. There are based on a model that takes amount of projected trends in demographics, capital investment and technological process.

Will the shift in global economic power be continued by technology process factor? They give these reasons to explain why technology

process can influence global economic growth as below:
The first reason, they feel future consumer online transaction market will trend mature that are becoming increasingly mature, sophisiticated and digitally technological process online transaction in popular. The second reason, emerging markets vary greatly in their institutional strengths and weaknesses and need to be assessed in a technological transaction way. These could also be major differences in institutional strengths between industry sectors within countries. Deep local knowledge that is updated in real time is critical here to manage businesses successfully in an emerging market environment, e.g. shorten online shopping transaction time and online product choice process. Having the right local partners to navigate online business through local political, legal and regulatory systems is also critical. Identifying and promoting local talent who understand local online business and social cultures better than any outside will also be an increasingly source of comparative advantage (As discussed in more detail in a recent PWC Growth Markets Centre Report, " Presence to prosperity"). The third reason, for larger western companies making strategic investments in emerging markets, part of their contribution could be to try to improve the local institutional framework. This could involve offering, appropriate technical assessment and advice to local governments in areas like corporate governance, fiscal policy and intellectual property rights protection. It could also involve investing in social and economic infrastructure (e.g. schools, roads, railways, power and water networks) where there are critical to a company's long term success in a region. Finally, there are existing markets in North America and Europe. There will remain very significant players in the global economy for decades to come. PWC report analysis shows average income levels will remain much higher than in even the best performing emerging markets for the foreseeable future. Advanced economics will also generally speaking, still be easier and lower risk places to do business given their political and institutional strengths. Thus, it seems that technological process will be the main

factor to influence global GDP growth and human living standard in our future societies. It also means that the economists predict the technological process will be one main factor remain to influence global economic growth and rasing human living standard.

Developing countries will face stronger headwinds in the decades and because technological changes are rendering manufacturing more capital and skill intensive. So, it brings this question: How past, present and future of human living standard is influenced by technological process development? To answer this question , we need to assume that economic growth is a precondition for the improvement of living standards and lifetime possibilities for the " average" citizen of the developing world. Optimists would point to improvements in governance and macroeconomic policy in developing countries and to the still not fully exploited political of global economic technological globalization to foster new industries in the poor regions of the world by outsourcing and technology transfer . Otherwise, pressimists would feel rich countries control world economy, threaten to technological globalization and obstacles that late industrializers have to surmount given competition from China and/or America owned high technolgical development etc. countries. Thus, it also brings this question: What will be changed to influence past, present and future economic growth and human living standard by technological development?

It is a global technology structural transformation change issue, it means that the birth and expansion of new (higher technological productivity change) industries and the transfer of labor from traditional or lower : productivity activities to modern ones. Thus, this history change will occur from past industrialization, then it will change to present technological process and it will change to future artificial technological industry in the future.

I. Has it close relationship between climate change and technological process to influence human living standard?

Some economists point out global growth will be hindered by rising operational costs as global temperatues rise, with studies suggesting that a worst-case impact of a 1% reduction in GDP growth per year could be realized. Research also suggests that the impact will be disproportionately damaging to developing economies and only through a collective effort to enact strict carbon emissions policies can be the long term financial challenge of climate change potentially. If it was truth, it will bring this question: Can climate change influence to reduce human living standard to be fallen?

To answer this question: We need to know why climate change can raise economic cost to global societies. Generally, we know global warming is likely to impact global activity. Such as, the effect on growth and inflation , climate damage functions: quantifying the impact on activity, regional poor effects. Accessing the impact of cimate change is complex with uncertainty about both the degree of future global warming and the subsequent impact an global activity. There are clearly some benefits as well as costs as the planet warms. There is also the unknown of how technological process will respond and potentially alter the path of global warming.

II. Why can climate change influence effect on growth and inflation?

Climate change at varying levels of warming, the impact of rising temperatures will be widespread, in part due to the financial, political and economic integration. Global warming will primarily influence economic growth through damage to property and lost productivity, mass migration and security threats. For Hurricane Sandy example, which flooded much of New York in 2012 year, it caused economic damage such extreme weather events can cause. Rising sea levels will also likely harm economic output and businsses become impaired and people suffer damage to their homes.

In economy view point, using a production function, we can demonstrate the likely effort climate change will have on output. If we assume less capital stock is available , due to the damage inflicted from climate change, we would see a fall in the productive capacity of the world economy. Then, the world products: function as each unit of labor produces less output, lower labor productivity may not just occur , due to a lower level of capital stock. However, higher global temperatures may affect food security, promote the spread of infectious diseases and impair those working outdoor. Such factors are likely to cause greater incapacity and social unrest and as a result will reduce both the effectiveness (productivity) and the amount of labor available to produce output. The argument whereby the world's population is seen not to respond to climate change. It is possible that, preventive measures , such as flood defenses are put in pace in order to avoid the costs of climate change. It is a short-term economic cost to this action as resources are directed from more productive uses. It can include two aspects influence , such as food inflation and energy cost increasing , the reasons are as below:

On the one hand, inflation is likely to rise as shortage, particularly in agriculture. Climate change will cause agriculture supply shortage and food demand rising. Due to a reduction in output, but an increase in the general price level as a result of global warming. This leads human onto the possible, an inflationary effects of global warming on the agriculture world economy. Because agricultural yields are sensitive to weather conditions and as our climate becomes ever more extreme, more frequent droughts may reduce crop yields in areas where food production is vital. Higher global food prices will likely thus causes the pressure to global consumers. Then, it will lead global food consumers' living of standard to be fallen down because we need to save more money to buy food. So our entertainment and daily essential spending expenditure will be also decreased, due to saving more money to buy inflationary food.

On the other hand, energy costs will increase, due to climate change influence natural resource supply number of energy production to be shortage. Thus, it will bring higher energy costs are also likely to boost inflation. As our climate become more extreme, we are likely to demand greater energy to both our working and living environments during the summer and heat , then when we experience cold winters. Thus, if human lacks high technological process development to prepare to win climate change challenge influence, e.g. invention of renewable energy or new food planting

or growing method which can adopt to grow or plant in bad climate in the future easily. Then, we shall face our living standard to be fallen , due to lacking technological development to solve climate change negative influence challenge in our future.
To conclude, I believe that human living standard will be fallen, due to we lack high technological process to prepare to solve climate change challenge, such as global warming, sea level rising, flooding, etc. natural crisis to cause food nature resource and energy nature resource supply shortage. Then, it will cause inflation of our essential product or food price rising seriously. Finally, as an effect, human living standard will be fallen.

Whether future technology can influence human living standard

I Can technology influence human habitable population dynamic change to impact living standard?
To explain above question, we also need to consider these questions, such as: What factors influence human population growth trends most strongly? How does population growth or decline impact the environment? Does urbanization threaten our quality of life or offer a pathway to better living condition?
Nowadays, human population trends are centrally important to environmental science because they help to determine the environmental impact of human activities. Rising population put increasing demands on natural resources, such as land, water and energy supplies. At the same time, pollution challenge will cause, due to human needs , spend or waste much natural resource to raise living standards, such as air and water pollution and greenhouse gas emissions, along with increasing qualities of water. Then, it will bring this question: Will technology influence environmental impact to human living standard?
Paul R. Ehrlich & John P. Holdren (1974) indicated one widely-cited formula is " I=PAT", which concerns population interacts with several other factors to determine a society's environmental impact as below:

Environmental impact = " population" x "affluence" (or consumption) x " technology"

In fact, the question depends on assumptions about human preference. What standard of living is seen as acceptable , and what levels of risk and variability in living conditions will people tolerate? Many of these issues are not just matters of what humans want, rather who intersect with physical limits, such as total arable land or the amount of energy available to do work. In such instances nature sets of bounds on human choices (Joel E. Cohen, 1995).

To research whether technology can influence human habitable population dynamic change to impact living standard. We need to answer this question. How did industrialization alter population growth rates so sharply? One central factor was the mechanization of agriculture, with enabled societies to produce more food from available in inputs.

In our past agricultural society, as food supplies expanded, average levels of nourishment rose and diseases declined over succeeding generations. Improvements in medical care and public health services, which took place more in uraban than in rural areas, also helped people to live longer. So , death rates face in our nowadays technological society. After several decades of lower mortality, people realized that they did not have. So, many families decided to born many children to achieve their desired family size as well as birth rates began to fall as well because human was educated popularly in our technological society. Thus, in our technological society, it will cause expanded work forces, it can help nations increase their economic output, raising living standards for everyone. They also can strain available resources and services. Anyway, it also may cause shortages and economic discuption.

However, I believe pollutants, medical services, mortality and fertility factors can influence human living standard to be better or poor, instead of technology factor. But, I also believe technology will be one important factor to influence human living standard. I shall indicate these reasons to explain why technology can influence human living standard to be better or worsen and it has effort to reduce these bad factors causing to influence our future living standard.

The first reason, exposure to pollutants is a major factor contributing to infant motality and lower life expectancy in developing countries, e.g. China, India, Hong Kong, Korea. Thus, if any these developing countries lack enough technological environmental investments, such as providing cleaner energy sources and upgrading sewage treatment systems which can significantly improve public health. So, these development countries‘ population living standard will have negative impact if they lack technological environment investments to reduce pollution serious level to influence citizens' unhealth.

In second reason, increasing life expectancy is creating a public health infrastructure that can identify and repond quickly to disease outbreaks, families and other threats. For example, when severe alute respiratory syndrome (SARS) emerged as a disease that might cause an international epidemic. The health centers for disease control and prevention launched an emergency response program that required health departments to report suspect cases for evaluation, developed tests to identify the SARS virus, and kept health care providers and the public informed about the status of the outbreak. The United States and many other countries also reported their SARS cases to the World Health Organization. These types of close surveillance and preventive steps to control infection. Thus, it seems nowadays, countries need to invest or medical technology on attempt to invent more unique medical machines to avoid any unknown diseases existing.

The third reason, that drives population trends migration when includes geographic population shifts within nations and across boards. Migration in less predictable over long periods than fertility or morality since it can happen in sudden waves, for example, when refugees brings a war. So, slowly over many years, immigration often changes host nations or regions ethics mixes and strain society. Immigration often changes host 's or regions' ethic mixes and social services.

On the position side, it can provide needed labor (both skilled and unskilled). In my view point, if the country has good technology resources to provide skilled immigration labor to attribute their technological reource is one important factor (production of factor) to let immigration skilled laboe to contribute their effort to serve its country to raise productivity to develop its economy growth for long term. For technological process source countries, however, technology can assist immigration to attribute valuable talent. Specially, since educated and motivated people (labors) are most likely to migrate in search of opportunities.

II Human's history technological development

In human history development, through the early decades of the industrial revoluation, life expectancies were how in Weatern, Europe and the United States. Due to lack high technological medical invention. Many people died from infactious diseases, such as typhoid and cholera, which spread rapidly in the crowded, bad environment conditions that were common in early factory towns and major cities or were weakened by poor nutrition, due to lack high technological environment science resources to provide environment protection professionals to produce any new technological environment protection product to reduce air and water pollution. But from about 1850 year through 1950 year , a cascade of health and safety advances radically improved living conditions in industrialized nations. Due to technological environment industry protection development. Some this technological environment protection industry development brings these benefits to human, includes: improving urban sanitation and waste removal, improving the quality of the water supply and expanding access to it, forming public health boards to detect illnesses and quarantine the sick, researching causes and means of transmission of infectious diseases, developing vaccines and antibiotics, adopting workplace safety laws and limits on child labor, and promoting nutrition through steps, such as fortifying milk, breads and cereals with vitamins.

By mid-20 th century, most industrialized nations had passed though the demographic transition, as health technologies mere transferred to developing nations, many of these countries entered the mortality transition and their population. The world's population growth rate peaked in the late 1960 year at just over 2% per year 12.5% in developing countries. Thus, it seems health death chance from diseases and environment pollution importantly. It implies our environment pollution to influence our health will be serious, so we need continue to develop my technology on health and environment protection aspects to raise human living standard to avoid or reduce death chance increasing.

III Can food technology production shift diets, the type, combination and quantity of foods people consume, contribute to a sustanable food future?

Building on the United Nations Food And Agriculture Organization's food demand projections, it estimates that the world needs to close a 70% "food gap" between the crop calories available in 2006 year and expected calorie demand in 2050 year. It indicates the food gap stems primarily from population growth and changing diets. The global population is projected to grow to nearly 10 billion people by 2050 year, with two thirds of those people by 2050 year, with two-thirds of those people projected to live in cities. In addition , at least 3 billion people are expected to join the global middle class by 2030 year. As nations urbanize and citizens became wealthier, people generally increase their calorie intake and the share of resource-intensive foods, such as meats and dairy in their diets. As the same time, technological advance, business and economic changes and government policies are transforming entire food chains from farm to fork.

In human future, the demand of increasing technological agricultural production will increase to efforts to reduce the food shorten to close the food gap. However, relying solely on increased production to close the gap would exert pressure to clear additon natural ecosystems. For example, to increase food production by 70% , when avoiding further expansion of harvested area, crop yields would need to grow one-third more quickly than who did during the Green Revolution. In short, yield increases alone with likely be insufficient to close the gap. In the fuure, farming technology development will be applied to solve these challenges. Such as human plan use the globagrimodel farming technology to quantify the land use and greenhouse gas consequences of different foods and then analyze the per person and global effects of the three diet shifts on agricultural land needs and greenhouse gas emissions. Moreover, scienists estimate to apply farming technology to these diet shifts, if implemented at a large scale, it can close the food gap by up to 30%. When substantially reducing agriculture's resource use and environmental impacts. Thus, it seems farming technology industry will be increased demand to applied to solve future food shortage challenges to raise human living standard in the future.

In conslusion, due to human's history development, it proved in our past, we had been encountering agricultural production to industrial production . Then in the present, we are encounting technological production. In the future , we will encounter human intelligence technological production in possible. So, it implies " future technology change will influence human living standard to be better if human continue consider to how to innovate medical health technology and agricultural growth technology and human intelligence technology and climate environmental protection technology and space technology research. Otherwise, human living standard can not be raised , even fallen if human is lazy and we do not continue to spend time and money investment to research these kinds of technological improvement. Thus, I think human ought continue to research these kinds of technological improvement to achieve human living standard to be better in the future.

Reference

Future of jobs survey, World Economic Forum.

Hauser, J. Tellis, G. J; Griffin, A. 2006. Research On Innovation: A Review And Agenda For Marketing Science. 25(6): 687-717.

J.P. Holdren & P.R. Enrlich, " Human population and the global environment", American Scientist, vol. 62 (1974), pp.282-92.

Joel E. Cohen, How many people can the earth support? (New York: Norton, 1995), pp. 212-36, 261-62.

Mohr, G. J. Griffin, A. 2010. Research On Innovation : A Review And Agenda For Marketing Science. 25 (6): 687-717.

Names of drivers have abbreviated to ensure legibility. Future of jobs survey, World Economic Forum.

" Presence to prosperity", PWC Growth Markets Centre Report: http://www.pwc.com/gx/en/growth-markets-centre/presence-to-profitability.jhtml

CHAPTER SIX

Technology how influences social development

Human Behavioral network job brings social economic benefits

What does human network job mean ? Why may human network job be popular? Why human network job behavior may influence economy ?

Nowadays internet is popular to use. We can apply internet to find data , search any new things, even earn money. Why does internet may become huma network job source. For example, e-publish may be one kind of new human network job. Any authors may apply internet

channel to help them to sell electronic or paper books from e-publisher web store. They may apply facebook, you tub etc. any online

channel to promote themselves new books to let new readers to know whether when they may buy themselves favourable new topic books to read

from electronic publisher web store.

Thus, future electronic publisher industry may help any authors to build internet network platform to help them to sell and promote ot advertise their any one new electronic or paper book topic to let global any one reader to choose to buy their any new topic books from electronic publisher web store easily and conveniently. However, it implies that electronic network platform author may be one kind of future new human network job in our societies.

How electronic network platform author job may bring economy benefit in macro economy view? A person can have few friends, contacts and still be very influential if these few
friends and contacts are themselves highly influential, e.g. one author must not need to know any one reader in global society. When they like to choose any electronic books from electronic internet network platform. They may become the author's any one topic book buyer, when they feel the author's any one topic book is fun and attract they make decision to buth the strange author whose the topic book from electronic book publisher's platform web store conventiently in short time. Although, they are strangers, they do not know themselves , but the reader can understand what it way that made Google from writing platofrm to create new creative mind and typing network job method to replace traditional hand writing book method for global authors. It will be one kind of new human network writing job.

Hence, global any one reader can apply an innovative search engine , such as google.com to find whether whom author personal new topic books are value to read from internet.
Then, the electroniuc publisher's web store may be new book store platform sale network to help the author to sell many electronic or paper books from electronic network platform
in short time. So, internet may be future new network plaform to help global any one author to create network writing job absolutely. Furthermore, internet may be popular social media
to help any one author to build goold relationship between his/her readers. It is one kind of new network, human network job. New authors do not need to buy many paper books to prepare to put in any one book shop warehouse. Their every book can print on demand to reduce out of book stock in any one book shop. They may choose to sell either electronic books or paper books both from any one book publisher web store. So, electronic network platform may be one kind of good writing channel to help human authors to create income and it can also help authors to bring new

creative mind and new topic fun content books to let readers to know and buy to read from electronic publisher network platform.

Why does human behavior may be one kind of new human network job to bring global economic advantages. ALthough, it may be free income or without inocme, but the person does the network behavior, his/her behavior may be bring advantages to influence many other people's health. For this case, when a worker in a coffee shop in an airport gets a vaccination aganinst the flu, it does not only helps him or her stay healthy, but also helps the many travellers who might otherwise have been inflected if that workers caught the flu. So, the externality , the result implies the vaccination of even a part of a community conveys benefits to the whole community. For example, governments pay special attention to the vaccinations of school children, teachers, health mothers, and the elderly, categories of people particularly susceptible not only to catching, but also to transmitting a disease.

It is not accidential that governments are heavily involved with vaccination . When there are externalities, free market, fail to persuade individual incentives with society's
their the worker's decision of whether to get a vaccine ends up attracting whether other people get sick. The workers might not fully take all these other people's potential suffering into account when making her or his vaccination decision.

As Stanford University does many suggestions, understand this and tries to help them make the right decisions and so providers free flu vaccines for its staff and students.

Small pockets of unvaccinated individuals can allow a disease to gain a spread more widely well-being. For example, parent weighing the costs and benefits of a vaccine for their child is not always thinking of the consequences of that vaccination to other people. THese are markets in which subsidizing or regulating behavior can make everyone better off. Because the reason for requiring that a child be vaccinated before enrolling in school is not just to protect that child, because each child's vaccination affects others via potential contagions.

Robots take our jobs behavioral and economy influences
Robot job behavior brings economy influences

If one day robots can replace human to do simple, even complex jobs. They will bring what influences to our global societial economy.The popular economic refrain declares that the global middle class is dying and robots will soon take our jobs, e.g. shopping center customer service jobs, library service jobs, cinema ticket sale jobs, restaurant kitchen cooker jobs, even, bus drivers, taxi drivers etc. public transport driving jobs, accountant, doctors etc. professional jobs. Whether it is beautiful or petty matter if our future societies have many human jobs can be replaced to do from robots. Businessman must may reduce to employ employees and reduce to pay salary or wage, when robots can be replaced to do their employees tasks. But, societies must bring unemployement rate rises , due to societies will have many people loss jobs when their employers choose to buy robots to serve their clients or do any office tasks or customer service or cleaning etc. tasks.

In micro economy view, employers may save money in long term, but in macro economy view, it will cause unemployment ratio rises , even crime rate rises when there are many people lose jobs in societies. These models of doom, though, fail to account for the hundreds of businesses riding the waves of change in their industries when robots may be invented to replace human to do many simple , even complex tasks in our future societies.

WE may image that one small factory needs to manufacture fishes canes to sell to supermarket, the small , cheaper stuff and higher margin parts of the fishes manufacture industry. Before, this factory needs to employe many human factory workers need to help every fresh customer makeing the perfect fishing gear, designed for performance, durability, and cost in order to achieve to manufacture every fish cane in whole fished processing manufacturing stages. Every worker needs to spend about 15 to twenty minutes to finish every fish cane , till to delivery to any

supermarket to sell. If this fish canes manufacturing factory can apply manufacturing robots to help them to finish any one working tasks , every robot can only spend five minutes to finish whole fresh fish cane manufacturing process. Thus, every robot can
help this factory save 10 to 15 minutes time to finsh every fish cane manufacturing process. IN fact, time is money, because when every robot can help this factory to reduce 10 to 15 minutes time to compare human worker. Then, this factory can finish about 20 fish canes in one hour if it can use robot to help it to manufacture fish canes. Otherwise, if this factory still use human workers to help it to manufacture fish canes, then it can finsh about 3 to 4 fish canes in one hour. SO, the manufacturing efficiency ensures that robots must help this fish manufacturing factory to raise fish canes number more than human workers. So, in robotic behavioral economy view, manufacturing robots must help this fish canes manufacturing factory to raise fish canes manufacturing number and deliver increasing number to supermarkets to prepare to sell every day. Robots can help this fish canes manufacturing factory bring manufacturing time saving, rising manufacturing efficiency, improving performance and reducing wages expenditure long time advantages in micro economy view. However, manufacturing robots can also bring disadvanages to society, e.g. increasing unemployment ratio, increasing crime rate,
this factory workers will lose jobs and income, they need earn social welfare from government and increasing government finance pressure in short time, even long time in macro economic view.

Stanford University graduate program in economics, Scott lecturer explained that "in demand and supply economic theory for robots supply and demand case, robots supply number increasing may influence human workers demand number decrease. It sometimes calls " the efficient frontier".
No specific human beings were mentioned in any of economics classes. As robots supply and demand in market case, They (robots) may be purely theoretical " agents" who reached to the most reasonable sale prices in order to persuade any one businessman

buyer to make manufacturing robot buying decision whether robots can help him / her to bring how much saving time , saving money, saving cost, improving performance, efficiency economic benefit before he/she plans to reduce workers number when he/she decides to apply robots to replace human workers in his/her factory or office or any service department, e.g. cinema ticket sale service, shopping center customer service, shopping center cleaning , supermarket customer service etc. service or sale tasks. When robots can replace human to do any one of these tasks in any organizations. So, robots may be human worker agents who reached to prices the way robots would react to a software
command. There was nothing that explained why some people thrived and others did n't or why truly brilliant, hardworking people could fail when much lazier folks succeeded." Having been admitted to the Stanford University graduate program in economics, Scott lecturer hoped to get his answers there.

How robots influence our future social changing? Using the right technology can be a boon to your business in this economy. For internet example, it is easier than ever to find well-matched customers all around the world, to stay in contact with them, and to more quickly design the products they want. If you focus solely on being cutting -edge, though you risk letting the technology
take over what should be very robust relationships with your customers , employees, and colleagues. IN nowaddays society, technoligical advances and cutomation, personal
relationships in business are more crucial than ever. I mean that robots can not replace human to serve clients to let them to feel more comfortable and passion more easily. For shoe shop case example, if the shoe shop apply one robot to serve its clients to replace human shoe salesperson to serve its shoe customers. Robots ensure that they can not persuade every shoe potential buyer to make shoe buying decision more easily when robots need to contact every shoe potential buyer. The reason is simple, because robots can not touch any one shoe buyer individual emotion very easier.
If the shoe buyer needs the robots to help him/her to choose any

right shoe styles when he/she can not feel himself / herself can make the most right shoe style choice decision. The robots can not replace human shoe salesperson to make shoe style choice judgement more easily. They must need longer time to analyze whether which shoe style may be the most suitable to the shoe buyer. Otherwise, human shoe salesperson may attempt to make the most right shoe style choice decision to help any one shoe buyer to chooce the most right style shoe because he/she owns shoe style sale experience, shoe style knowledge, the most important reason is that they can feel every shoe customer individual emotion to touch whether he/she will feel comfortable or happy when they attempt to help every shoe customer to seek the most right shoe style in every shoe customer whole shoe searching processing. Othwerwise, serving robots are only one machine, they can not touch or feel every shoe customer individual emotion whether he/she feel comfortable or unhappy or happy when they need to contact them in whole shoe searching processing. Hence, I believe that some tasks robots can

not repalce human staff to do very easily. Otherwise, robots may bring disadvanatges to let any one businessman to loss his/her customers, due to robots can not touch every customer

emotion to compare human staff in service tasks more easily. Robots serving customer behaviors may cause money lose and customers number lose to the shop in micro economic view.

Intellectual human economic behaviors

What does intellectual human economic behaviors mean ? I believe that when we choose or decide to do intellectual behaviors, then our societies will be influenced to bring economic growth in consequence.I shall attempt to indicate pollution case to explain how and why eithet our intellectual or foolish behaviors may bring economic growth or recession in consequence as below:

On one hand, for air pollution social case aspect example, if we only consider to buy cars to drive for working aimr or holiday leisure aim. Then, our societies air will be polluted. Our health will be influenced to bad. Our car driving behaviors may cause global environment air pollution serously. In long tiem, global air pollution will bring our bodies health to be bad. Although, ourselves car driving behaviors may bring our driving travelling leisure enjoyment and comfortable feeling in short time, also we so not need to pay public transport fare often, but we need to compensate ourselves health economic intangible loss due to air pollution , when cars number increases, dirty air will cause ouselves health to become bad.

In the result, we will need to pay more medical expenditure when we are old age, due to ourselves bodies will become bad, due to we breathe global dirty air every day, due to ourselves cars pollute air in long time, e.g. 10 to 20 years, even 30 more without limited air pollution environment. So, driving cars behavior may be one kind of human foolish behavior and our foolish behavior may bring ourselves future long time medical expenditure absolutely.

One the other hand, water pollution social aspect, if we often keep much rubblish to pollute sea, oil exploration porcessing pollute ocean , ships gas pollute ocaen, then fishes will eat polluted food and drive dirty water, due to global ocean is polluted.

In fact, because human only to conside how to buy boats to carry on leisure enjoyment activities, or catch cruises to travel on the sea. Also, oil manufacturers only consider researching anywhere to find new oil exploration places to manufacture oil product, when their oil exploration processes pollute ocarn . Consequently, global fishes drink polluted warer or eat polluted food. They will have poison. SO, human will have high chance to eat poison polluted fishes, due to fishes are poison or are polluted.

So, human is doing foolish activities, we only hope to find oil exploration places to pollute ocean or we only spend money to buy ticket to catch ships to travel anywhere in global ocean. All of these human foolish behaviors will bring pollution to global ocean. On consequently, we will need to compensate to eat polluted or dirty or poision fishes, ourselves bodies health will be bad. In long time, we need have high chance to pay medical expenditure when we are old. So, pollution case may be one good example to explain how and why human foolish behavior may influence ourselves future need to compensate serious medical loss.

All of these human foolish behavior will bring pollution to global ocean. On consequently, we will need to compensate to eat polluted or dirty or poison fished , ourselves bodies health will be bad. In long time, we will have high chance to pay medical expenditure, when we are old. So, pollution case may be one good example to explain how and why human ourselves intellectual or foolish behaviors may influence future long time economic loss or economic growth or recession in micro and micro economic view.

On another water pollution aspect hand, if we often keep rubbish to sea, oil exploration processing pollutes ocean and ships' gas pollute ocean, then fishes will eat polluted food and drink dirty water, due to fishes will eat polluted food and drink dirty sea water because the global ocean is polluted seriously.

In fact, because human only consider how to buy boats to carry on any leisure water activities, or catches cruises to travel on the sea. Also, oil manufacturers only consider any where to find oil exploratin places to manufacture oil products from ocean, when their pol exploration processes can plooute ocean. Consequently, global fishes drink polluted water or eat direty food. They will have poison. So, human will have high chance to eat poison fishes.

Otherwise, such as pollutin case, it can infuence inflation or deflation. Consequently, the reason indicates supply and demand theory. If air pollution is serious, then we will consider health issue, global cars demand number may be influenced to reduce, when global cars number demand will reduce, global car prices and supply number will need to change to fall down in order to attract or persuade global car consumers choose to make car purchase decision.

Hence, global car manufacture number and car price will be influenced to reduce, due to global air pollution issue. Consequently, deflation will occur because when the country citizen usually does not spend much extra saving money to buy car expensive goods. Money value will be low. Otherwise, if global cair pollution is not serious, human considers to buy cars to enjoy driving leisure lives. So, global car demand is influenced to increase , also global car price will also influenced to increase.

Consequently, gobal human will choose to buy cars to drive. Due to we accept to spend extra saving to buy expensive car goods. Car sale price and supply may be influenced to rise up. Money value is influenced to reduce. Inflation may be influenced, due to global car consumers number increases, we would not have extra money to spend easily. Car expensive goods expenditure influences our spending habit to avoid to make car purchase decision more easily. So, human intellectual or foolish activities may bring inflation or deflation consequency in possible indirectly in macro economic view.

On conclusion, above pollution case explain that how and why human intellectual or foolish economic behaviors may bring inflation or deflation consequency as wll as economic growth or recession consequency as well as any goods demand and supply increasing or decreasing consequency. It implies that human behavior may have indirect relationship to influence any goods demand and supply number to either increase or decrease result as well as any goods price will be influenced to increase or decrease in micro and macro economic view.

The relationship between social change and human behavior

Why does economic changes may influence human individual behavioral change? I shall attempt to indicate shopping behavior and staying at home behavior to explain their case and effect relationsip as below:

Human behavior can be influenced by economic change or economic change can be influenced by human behavior? Why does recession may influence consumers reduce shopping desire? In social recession suitation, it is possible that many people lose jobs suddenly, due to businessmen lose many customers. They need to make decision to reduce employees number in order to continue to keep businesses. Consequently, many firms (organizations) their employees may lose jobs. When they have much time, due to lose jobs, they will feel to avoid to spend too much time and money to go to shopping often. Many losing jobs people, they will often stay at homes.

So, they will reduce time to go to shopping, then non essential products won't their preferable choice purchase products. Hence, recession will change many losing jobs people their shopping or consumption desires to avoid to buy non essential products often . Usually when economic boom, many people have jobs to do because consumers number must increase when many people have jobs to do. Then, many people can accept to spend money to buy non essential products often. Many people feel spend time to go to shopping can satisfy their purchase of any kinds of new products useful psychology or desire. So, recession is one good example to explain it can influence many people do not like often to leave homes to go to shopping easily. Many people like to stay at homes, becaue they feel worry about spending too much shopping time when they leave homes. Their staying home time is one good negative shopping behavior example. So, economic change may influence human individual behavior changes , they have direct cause and efect relationship in behavioral economic view.

May human behavior influence economic change? Is it possible that human behavior may bring the country social economic change in macro economic or micro behavioral economic view ? I shall indicate publishing industry example. Do you feel that if there are many students feel learning is very important when they read many books or many of students feel interesting to read or they have reading new books in habit, then it is possible that the country will have many students like to spend time to go to any book shops to choose the books, they feel that they can help they learn new knowledge. Then the country will increase students number, they often spend time to visit any one book shop every week. Their visiting book shops behavior which may become their habits. So, the country will increase students number, they often spend time to visit book shops. Also, it implies that visiting book shops behaviors may be their behavioral habits.

So, when the country has many students often spend time to visit book shops , their visiting book shops behaviors may help any one book shop to raise books sale chance. So, the country's student individual often visiting book shop behaviors, their habitual visiting book shops behaviors must may assist help any one book shop to increase books sale number absolutely.

Consequently, any one book shop , its books sale bumber must be influenced to increase to increase because the country will have many students like or feel need visit book shops habit in order to choose any suitable books to buy to read at home in order to raise themselves learning effort. When the country has many bok shops often have many students visit their book shops, then their books sale number may be influenced to increase. It explain why student individual visiting book shop behavior may help any one book shop sale number increases also.

How human productive behavior may influence economic development

May any country which citizen behavior assist themselves country development? It is one cause and effect economic question. I mean that if the country itself citicen can not concentrate mind or energy to choose to do one kind of industry in order to let themselves country can bring the most benefit, then whether the counry itself economy can bring the most serious economic benefit. I shall attempt to indicate these countries themselves indistry choice to explain whether these countries themselves citizen productive behavior may help themselves countries to achieve the largest economic benefits. I shall indicate as below:

New Zealand farmer individual wine productive behavior

For New Zealand country example, this country concerns itself effort is foucs on farming agricultural aspect. So, this country has many farmers concentrate on farming agricultural aspect. May New Zealanders choose to spend time to produce different kinds of wines, e.g. wine or red grape wine is for the people are eating meat, or they are eating dinner.

When these New Zealanders their behaviors choose to do farming or agriculture to grow and produce different kinds of taste of white or red grape wine drinking products job. Themselves grape agriculture behavior will influence these New Zealanders themselves, they can learn how to improve different kinds of grape wine drinking products in order to achieve every kinds of white or read grape wines taste improving aim during their white or red grape producing process.

Why can New Zealander every individual white or read grape wine producers improve their white or read grape wine taste more easily? In behavioral economic view, it can explain that why any one New Zealander white or read grape wine producer can be encouraged or excited or persuaded to concentrate nervous and energy and effort to learn how to improve their white or red grape wine products easily.

In fact, New Zealand is one agricultural food export country. It has good natural environment resource , e.g. land, seed to provide any one farmer to produce themselves any kinds of agricultrual food products, e.g. fruit, or wine food products. Because New Zealanders know themselves country has enough natural resource . So, in common, many New Zealanders choose to attempt to do farming agricultural jobs in order to export themselves any kinds of fruit or meat or wine products to overseas or sell to domestic in order to earn profit.

So, when these New Zealand farmers number has been increasing every year. This country farmers will feel themsleves competition between this New Zealand farmers themselves are serious due to they may feel New Zealanders choose to do agriculture businesses in order to export themselves different kinds of farming food to overseas or sell to local to earn profit.

Hence, when many New Zealand farmers feel that farmers number has been increasing every year. They will feel themselves competition is serious. They must need to spend much time and nervous and effort to research what method is the best how to produce the best taste of white or red grape wine products in order to let local or overseas wine buyers to choose to buy his/her

producing white or read grpae products to drink.
Hence, in competition psychological view, may influence many New Zealand white or reaad wine producers had been beginning to change their learning behavior on researching what method is the best in order to produce the best quality of taste red or white wine products to sell in order to attract overseas or local white or read grape wine drinkers to choose to buy his/her wine products. Their behavior will focus on learning how to raising or improving white or read grape wine taste method more than only focus on producing a large number white or red grape wine products. They believe wine quality is more important to compare wine producing number. So, New Zealand wine producers themselves wine producers behaviors have been changing on concentrating on researching wine quality method aspect more then wine producing number aspect in behavioral economic view.

America high technological productive behavior

For America example, US is one high technological country, it owns many high technological knowledge talent inventors, e.g. computer science inventors. Hence, US must attract many diferent countries owning high technological computer inventors choose to go to US to develop their computer science profession career. Also, it seems that when many computer science inventors or professions choose to go to US to develop themselves computer science new career. In behavioral economic view, due to their leaving themselves countries choice, which may bring influence themselve country job behaviors need to be changed. They must need to adapt US new live. Because they will forgive their past computer science job. These computer science professionals need to spend time to adapt US new lives. They " past computer science job behaviors" will need to be changed to their new US any computer employer's new computer science job model.
Because their traditional computer science jobs needed to be forgot in their themselves countries. They will feel their old computer science job knowledge and behavior needed to change in order to let their US any one new of computer company employer feels

satisfactory to accept their new working behavior in any one US computer organization.

So, on the other hand, many US computer company employer will feel that they must need time to accept any one new overseas computer science professions their working behaviors, their working attitude daily, because these foreign comouter science professional, their past computer working behaviors and working attitude must be different to US domestic computer science professions.

In behavioral economic view, these overseas computer science professions, their working behaviors and attitude must be needed to change in order to adapt any one US new computer company itself domestic or local computer science professional stafs themselves daily working behaviors and attitude because these overseas and local computer science professionals must need to team work together.

In behavioral economic view, it is only one way that foreign computer science professionals must need to change themselves past country traditiona daily working behaviors and attitude in order to cooperate with these US local computer science professionals in teams more easily.

Consequently, if these foreign compute science professionals can change their past working behaviors and attitude to let any one US local computer science professional feels to cooperate with them easily in short time. Then, the US computer company itself whole computer professional teams themselves efficiencies will be influenced to raised or improved by the changing past working attitude and working behaviors of these foreign computer science professionals. So, in behavioral economic view, only if US any one computer company hopes itself computer teams themselves efficiency can be raised or improved when it decides to employ foreign computer science professionals and US domestic computer science professionals. They need to work in teams together. They must need to let these foreign computer science professionals to know how to change their working behaviors and attitude to let their domestic computer science professionals feel easy to work together. Then, the US computer company itself whole team efficiency must be rasied or improved easily in short time.

● China share market investing behavior

For China share market example, economic development depends on financial market. Because if many Chinese have interest to invest to carry on shares buying and selling activities in orde to learn how to earn shares interest and share profit when the China shareholder can make decision to sell himself/herself shares in the the high price, then he/she can earn money when he/she can sell the China company's shares in the high sale share price position.

If China has many Chinese like to spend time to carry on investing shares activities. Themselves shares buying and selling behaviors will influence China has many companies can increase fund from many Chinese shareholders in order to have enough money to expand or develop themselves businesses in China in long term.

Consequently, when China can have many Chinese like to attempt to carry on buying and selling shares investing behaviors in China share market. Themselves buying and selling shares behaviors can help many Chinese companies have effort to increase enough money or capital in order to continue to do their businesses in long term absolutely. So, it explains why when many Chinese become shareholders , they can assist China will have many companies continue to develop their businesses if many Chinese like to carry on shares buying and selling investing behaviors in long time in China financial investment market nowadays in behavioral economic view.

Why has any individual country have many people invest share behavior which can influence the country's macro consumption desire?

I shall apply shares market buying and selling investment behavior to explaiin why shares investment behavior which may impact the country's overal consumption desire as below:

In behavioral economic view, I assume that when the coutry has many people have interest to attempt to carry on shares buying and selling investment behavior, then their frequent shares buying and selling behaviors which may bring negactive consumption desire or shopping desire of these shares investors their consumer behavior. The reason is simple, when the country has many share buyers number suddenly been increasing rapidly. Consequently, these large group share investors must need to spend much time to research any kinds of company shares variations, whether when their share prices will rise up of fall down in order to achieve buying the company's shares in the lowest price and selling the company's shares in the highest price level in order to earn profit.

Basic on this reason, they must need to spend much extra time to research share prices changing behavior every day, e.g. one working person will wait to leave his/her job, after he/she can spend time to gather data to research the day's share price changing behavior after dinner. So, the working person's right time may be his/her share price market research behavior. Before he/she may spend his/her night time to go to shopping after dinner, but nowadays, he/she will fogive to do his/her shopping behavior before dinner or after dinner at hight sometime. He/she will make decision to spend much night time to turn on computer to click on share market website to research his/her share purchase choice to investigate whether his/her share price whether it rises up or falls down at the moment in order to make his/her share buying or selling decision at ever night time.

I mean the when the country has many people are share investors, their shares investment behavioral spenging time which will influence many shops lose customers at might often because the country will have many people feel need to spend night time to turn on computer or watch television to investigate share price variation. So, the country will have many people / share investors choose to stay at home in order to carry on share price variation investigation behavior, they need to listen share market update news from radios or watch the share market update news from computer or TV at home every night. Consequenly, they must reduce times to leave themselves homes at night. So, their shopping behavior also will be reduced. Because these share investors feel need to spend time to investigate share price variation news at homes which can bring economic benefits (high opportunity benefits) when they choose to forgive to leave homes to go to shopping times (opportunity cost) every night.

On conclusion, it seems that when the country has many people are share investors, then their share price investigating behavior may bring negative shopping emotion at night. Consequently, the country's any one shop may lose many customers from this share investor consumer group in behavioral economic view. Hence, when the country's share investors number had been increasing rapidly, it will influence any shops lose many customers from this share investing customer group at night frequenly in short time, even long time in behavioral economic view, because their shopping desires or shopping emotion will be brought negative feeling when they make decisions to spend much time to listen radios or watch TV or computers share price update nes at night. Hence, share market will bring negative impact to influence consumer shopping desire or negative shopping emotion in behavioral economic view.

Can technology influence human shopping behavioral change?

Nowadays, technological development has reached mature stage, whether technological mature stage may bring positive or negative shopping emotion influence to global consumers. I shall aplly internet inventin or ecommerce shopping channel tool to explain whether internet technology can bring postive or negative influence to global consumer behavior in behavioral economic view.

Internet is a good technological tool, it brings e-commerce business chance. In fact, commonly, global has have many businessmen choose to use internet channel to carry on their products transactions between global online-buyers and their electronic websites. So, global many shoppers had begun to feel online shopping is more convenient to compare visiting shops shopping.

Their shopping behaviors have been changed from internet technological tool. Global has many shoppers choose to buy any products from any overseas or local businessmen their web stores. They only need to spend time to find any businessmen their webstores to choose the most suitable products to pay visa to buy from their webstores. at homes. So, in general, global had have may shoppers had changed their shopping behaviors from visiting shops to visiting webstores at homes often.

So, it seems that internet technological tool had influenced global many shops disappear, but internet webstores will be replaced their actual shops on streets. Some of businessmen either they choose webstores to replace shops or choose websotes and shops both or still keep shops only. Hence, internet tool influences global businessmen have three kinds of products sale channels to let globa local and overseas consumers to choose how to buy their products. However, in fact, many of global shoppers, youngers and olders had begun to accept to buy any products from webstores. They feel to spend time to leave homes to visit shops , their shopping behaviors will be wasted time to not essential part to their daily lives. Hence, since internet technological invention, it had changed many consumers their traditional visiting shops shopping habit to change to buying products from webstores channel.

However, on the one hand, internet creates webstores ecommerce shopping channel to let global many consumers do not need to leave homes to go to shopping. It brings negative visiting shops shopping emotion to global general consumers nowadays. But on the other hand, it also brings positive visiting internet webstores shopping emotion to global general consumer nowadays. So, it seems that global many consumers feel that they often do not need to spend much time to go out shopping. Many global consumers feel convenient and enjoy to choose any products to buy from different internet webstores, when the online buyer chooses the most suitable product, he she only needs to pay visa card to buy the product from the online seller's webstore conveniently at home.

Hence, online shopping can bring economic benefit to online

buyers, e.g. avoiding walking time or spending transport fare to visit the shop to go to shopping, shortening or reducing shopping time to do another important matter.

On conclusion, global many consumers began feel online shopping can bring more economic benefits on shortening shopping time, avoiding transport fare spending aspect. So, online shopping will be popular shopping behavior for future long time. It may encourage global many shoppers can make rapid shopping decision in short time in order to carry on any products buying transaction to global any one online shopper in short time easily in behavioral economic view. So, global many businessmen had begun to build themselves one attraction webstore in order to persuade different countries consumers to choose to click themselves webstores from internet channel to buy any kinds of products in short time easily.

So, internet technology had changed consumers traditional shopping behaviors to build positive online shopping emotion as well as raise online sellers' any products sale chance easily in behavioral economic view.

Why and how human behavior may influence the country's economic growth or recession?

When one country has many people choose to do the same matter for one period, whether their behavior may influence the country's pvera; economic growth or recession . I shall attempt to indicate cases toexplain their relationship as below:

For flowing rubblish behavioral case example, do you feel that when the country has many people often flow rubblish on the streets, instead of their flowing rubblish behavior may bring streets dirty? But, their flowing rubblish behavior may explain that this country has people may have enough money to buy food to ear, or enough cloths to wear, enough bottles of water to drink, even they may have enough money to buy new television, radio, refrigeraters , washing machines, desktops or laptops electronic home products from old to new to use in order to satisfy their living needs. So, when they flow old electronic home products, their flowing old home electronic products behaviors may seem that they have enough money to buy other new home electronic products to replace old home electronic products to use at homes.

However, it seems thaat this country ought have many people have jobs to do. So, many of them, they can easy to make purchase decison to flow any old home electronic products and buy any new home electronic products to use . Because this country has many people have jobs to do. So, they can often not use old home electonic products to become rubblishs to flow on streets after they had bought any kinds of new home electronic homes.

In fact, it also implies that this country's economy grows rapidly. So, many businesses can glow up rapdly. When they expanded their businesses, they must need to increase employees number in order to let they help themselves to raise productivity or serve their clients absolutely. So, when the country has many businesses can grow up, it seems that its economy must be better or it is improved to compare past. Due to many different kinds of home electronic products had been often bought to use by this country people in this period. So, this country's any streets can be observed that expensive electronic home products were flowed on streets anywhere. then, this country will have many electronic home products sellers can sell their home electronic products very easily. When this country has many people can find any kinds of jobs to do easily. So, due to unemploymen rate had been decreasing.

In behavioral economic view, as this many electronic home products rubblish country case, we can observe this country may have many people have jobs to do. So, consumption number has been increased long time. So, cheap food, or expensive home electronic products may be rubblish on any streets. This country's people , their flowing rubblish behaviors may be explained that many of people have enough jobs to do, so they have ability to buy any good taste food to eat or buy any kinds of expensive electronic home products to use. So, this country's economy may be improved for this long period. So, in behavioral economic view, when this country can have many electronic home products rubblishs are flowed on anywherer in streets frequently. It seems that this

country will have many people have jobs to do, so it causes they often change old home electronic products or replaced them easily, when they have enough income to spend to buy any kinds of new home electronic products to use at homes easily. Moreover, their flowing old electronic home products behaviors also indicate that this country has many people their salaries may be increased in possible from their emplyers. When this country can have many different kinds of home electornic products are sold. It means that this country's electronic home products needs or demand had been increasing, due to many people have jobs to do and income increases to excite their living of needs also improve. Consequently, this country may seem have better economic improvement. We can observe from this country's electronic home products rubblish increasing income in theis period.
On conclusion, this country ought experience economic growth at this period. So, " flowing expensive electronic home rubblish increasing number " may seem that this country's economic growth is rapidly in this period, due to many people have jobs to do as well as salaries increase in this period.

Technology how impacts human behavior changing?
Technology how influences human behavior to bring changing? For example, online share purchase and sale transaction from smart phone brings share investor can do share buying or selling transation in any where and any time conveniently, non manual driving auto vehicle, bring car owner feels comfortable and spends free time to do other matter, e.g. reading, listening mucis in himself or herself car freely. electrical energy vehicle can help car owner to reduce air polluton and it can brings the drivers do not feel drive long time in any journeys in order to avoid air pollution for environmental protection responsible car drivers in our societies. Thus, they will drive long time in any journeys when they can drive electronic energy cars to replace oil energy cars.
However, online technology can also bring consumers can choose to stay at homes to buy any things from seller individual online

webstore conveniently. Such as online technology can bring shoppers do not need to spend much time to visit shops to buy any things. They can choose any kinds of products from any online sellers individual online webstores conveniently at homes. Online technology excite busy consumers can make purchase decision easily as well as it can help online sellers sell any kinds of products from internet easily.

In behavioral economic view, technology can change human behavior to be improved, it can let human feels comfortable, more free time ro use, rapid making any decisions, such as apply smart phones to make share purchase or sale transaction decision, online shopping decision, even travelling any where decision in short time, when the traveller finds the most cheap hotel accommodation room price and air ticket price frm any travel agent online tourism webstore, then the potential travel customer can follow the online hotel accommodation price and air ticket price data to make decision when to buy the air ticket from the airline travel agent or make decision when to prebook which hotel accommodation room to go to the country to travel from online travel agent tourism webstores. So, technology can encourage global any country travelers to make anywhere to trvel rapidly. If the traveler can find the country's general hotel rooms and airline tickets prices had been decreasing more sightly. The traveler may make travel decision to choose the country to travel in short time, then he/she can prebook the country;s any hotel room and airline ticket to pay by visa fraom the country's any hotel and airline travel agent webstores., before one week, even one month or more easily. Hence, online technology can also encourage traveler individual frequent travel times to be increased, due to global travelers can find any hotel rooms and airline tickets prices from internet conveniently at homes. They do not need to spend time to visit any airline travel agent to enquire travel choice country's hotel rooms prices and airline ticket prices. They can compare global travel of countries choices ' all hotels rooms and airline agents air tickets prices to make prebook airline seat and hotel room decision before

one week, one month even six months early.
On conclusion, online technology can encourage global travelers can make travelling any where and when traveling time desicions easily. It can excite tourism industry develops in long time. Also, such as electricity cars invention can encourage environment protection car owners do car purchase decision easily, because they can choose to drive electronic energy cars to replace oil energy cars in order to avoid air pollution occurs easily. So, electronic cars can increase electronic car purchasrs number, due to many of environmental protection attitude of car owners can choose to drive electricity cars to bring air cleans, even non -manual driving cars can encourage lazy driving and free time driving car owners to choose to buy non-manual (artificial intelligent) cars to drive , because they can spend much free time to read, listen music or do any matters in themselves cars, they do not need to drive cars, robotic (AI) auto driving machine is such one non-manual driver to help them to drive themselves cars confidently. So, non-manual driving cars can attract lazy and enjoying free time driving car owners to choose to buy to replace traditional manual cars to drive easily. Moreover, online share transaction can help any share investors to make share buying and selling decision in short time easily. When they can apply smart phones technological tool to carry on share buying and selling activities easily. They can observe any share rising or falling price suitation from smart phones in any where any any time easily. So, smart phone technology can help global any shareholders to make share purchase and sale transaction easily. So, technology can encourage human makes decision in short time rapidly.

How and why employees behaviors may influence economy development?

In behavioral economy view,I believe the country's any organizational employees behavior may bring indirect relationship to influence the country's long term economic development. I shall indicate past manufacture industry social development period to explain their relationship. For many countries' past business activities had belonged to manufacturing industry, such as US, UK past before 1980 year, it focused on steel manufacturing and steel manufacturing related machine products. So, US, Uk developed countries manufacturing industries may be past main country's economic income sources. I assume US , UK past had one million number different kinds of industries. They ought had about seven houndred thousand number organizational businesses were belonged to manufactured industry. They may include:

Steel manufacturing and steel related machine manufacturing, e.g. vehicle manufacturing, home appliances, e.g. washing machine, television, radio, refrigerate cooler, heater, air condition etc. different kinds of different kinds of steel -related manufacturing machine, they were manufactured from US, UK steel machine manufacturers. So, US, Uk the other three hundred thousand number industry may be general service industry, e.g. hotel service, restaurent, cinema, public transport service, tourism lesiure , wine bar, supermarket etc. different kinds of non-manufacturing industries business organizations were operated in UK, US past before 1980 year.

So, in UK, US developed countries industry development history, they ought have high percentage of businesses belonged to steel related manufacturing machine and steel products. Also, in the past before 1980 year, US, Uk business employers , they employed many workers are manufacturing workers. They needed to spend long time to work in factories. They were skillful workers, and they are trained to manufacturing cars, washing machine, television, heater, etc. even steel itself different kinds of steel related products to prepare to deliver to their shops to sell to US, Uk local or overseas clients.

So, I believe that past UK, US ought employ many employees, they belonged to skillful manufacturing workers, manufacture increasing steel machine or steel related machine number of products rapidly daily. So, if UK, US had had many of these manufacturing factories owned high skillful workers, then their manufacturing steel-related machine or steel both kinds of products number must be influenced to raise rapidly. Consequently, their steel machine manufacturing products would been exported to overseas or would been sold to local both markets , they may be influenced to raise sale number. They (these manufacturing workers) needed to be trained to know how to manufactur these different kinds of machine products in the efficient teams and they ought to be trained to raise their efficiencies in order to shorten time to manufacturing many kinds of steel related manufacturing machine or steel itself products rapidly. So , if their efficiencies and manufacturing performance was improved, these US, UK any one manufacturing worker and their teams ought achieve raising productivities significantly.

Hence, when past UK, US manufacturing industry development period, if these two countries' any manufacturing factories could have many manufacturing workers could be trained to be skillful and proficient manufacturing workers. Then, in past every day to these factories workers, they ought help their steel or steel related manufacturing employers to raise any kinds of machine or steel products number in every team. So, when past in the manufacturing industry development, US, UK could have many factories' manufacturing workers themselves steel or steel related machine products manufacturing skill could be trained to to improve to any kinds of these machine or steel manufacuring products quality as well as their products number could be influenced to raise by themselves skillful improvement significantly every day.

Then, what would be influenced to occur to past UK, US manufacturing industry period? In behavioral economic view, when these two manufacturing industry developed countries, such as UK, US , if they had many factories workers can be trained to improve their skill in order to achieve any kinds of steel or steel-related machine products quality could be improved as well as products manufacturing number could be also increased absolutely. In consequence, past UK and US both countries ought increase themselves any kinds of steel and steel related machine products number to be supplied to themselves local shops to let local clients to choose any one kind of machine manufacturing products to buy easily as well as they could also export to supply overseas any countries to buy their different kinds of steel or steel related machine products to let overseas steel or steel related manufacturing machine product buyers, they can have many of these different kinds of these steel or steel-related different kinds of manufacturing machine from UK and UK these both countries easily to compare other countries.

On conclusion, I believe that past US, and UK macro manufacturing industry income GDP would increase significantly. So, they would have good economic growth performance because when many of these manufacturing workers themselves manufacturing effort could be improved. So, it explained when employees manufacturing abilities can influence economic growth indirectly.

Robots invention whether they can help organizations to raise efficiencies or inefficiencies?

In behavioral economic view, in any organizations, when the organization hopes its worker teams can raise efficiencies , the organization may choose to increase more workers number and/or it can provide training to improve these workets themselves skills in order to raise their efficiencies. For one warehouse example, when the warehouse increases many goods , they are needed to delivered these goods from the shelves to the delivering destination locations. If this warehouse supervisors feel these workers themselves goods delivery speeds are slow, which is possible due to this warehouse's workers number is not enough. So, this warehouse supervisor ought increase workers number in order to increase their goods delivery speed in order to deliver goods from the shelves to every indicated goods delivery destination in order to let any one lorry driver can transport the right kinds of goods and ensure the accurate goods number to transport to any one client home rapidly. However, if this warehouse supervisor planed to buy several warehouse goods delivery robots to assist these warehouse workers to find the right kinds of goods from shelves and then deliver to the right destination location in the warehouse. So, these warehouse orkers can concentrate on counting the accurate goods number and ensuring the right kinds of goods in order to prepare to let lorry drivers to transport these goods to these goods of buyers themselvers homes rapidly. Consequently, in the first step, robots can concentrate on finding th right goods from shelves and delivers them to the right goods transportation of location destination. Then, in the second step, these warehouse workers can concentrate on counting the accurate goods number and ensuring the right kinds of goods in order to prepare to put them to the lorry. Consequently, when warehouse robots and warehouse workers can cooperate to work together, the most important, robots, can deal on finding the right kinds of goods and deal on delivering the accurate number of goods of job duty as well as these warehouse workers can only concentrte on counting the right kinds of goods number in order to avoid it has none any mistake of wrong kinds of goods and inaccurate goods of delivery number to be transported to the lorry and to deliver to any one buyer's home.

So, it seems that warehouse robots ought help any one warehouse worker to raise himself efficiency and avoid goods delivery of mistake occurrence easily as well as their help to warehouse workers that can let any one goods buyer feels their goods can be delivered to their homes rapidly. Moreover, warehouse robots can also help these warehouse workers to raise efficiencies because warehouse robots can help them to shorten goods delivery time between any one shelf and any one goods delivery destination of location in the warehuse because robots may help them to find the right kinds of goods from the right shelf in the short time. So, any one worker does not need to spend long time to seek anywhere is the right shelf location for the kind of goods when the kind of goods are needed to deliver to the buyer's home from lorry. Warehouse robots can help them to do this aspect of " finding the goods from the right shelf in short time job duty". So, any one warehouse worker only needed tospend less time to do the counting of any right kind of goods number and ensuring the right kind of goods job duty. Consequently, this warehouse 's any one worker, his any

one kind of goods delivery time may be reduced, because robots' assistance and they may have more confidence to avoid mistake to deliver the wrong number of goods and/or the wrong kind of goods to any one goods buyer's home.
On conclusion, it seems that warehouse robots ought may help any one warehouse worker to raise efficiency for any one team in the warehouse as well as the warehouse any one supervisor does not need to spend much time to observe any one worker individual performance for " goods delivery job duty aspect" because their goods delivery job duty that had been replaced to do by these several warehouse robots. Robots can achieve the more accurate of right kinds of goods and the right number of goods delviery job performance to compare any one of human warehouse worker themselves right kinds of goods of delivery and right number of goods of delivery job performance. So, when robots can participate to cooperate with this warehouse's any one worker to do their goods of delivery job duty in this warehouse every day. Then, robots can raies any one of supervisor individual confidence in order to let they do not need to spend time to observe any one of worker individual whose goods of delivery job performane. They can concentrate on supervising any one worker whose goods transport to lorry in the final step in order to avoid to deliver wrong goods number and / or wrong kind of goods to any one goods buyer's home every day. Consequently, this warehouse's overall teams of their delviery of goods performance many be improved by robotss' participatin to goods of delivery task as well as this warehouse's oveall teams themselves efficiencies may be influenced to raise by robots' goods of delivery task participation.

Why social behavior may influence organizational strategy needs to be changed ?
Why any organizations need to know whether nowadays social behaivor how has been changing in order to implement the kind of the most right strategy to achieve the profit aim pursue in possible. I shall indicate nowadays ecommerce or online, customer shopping

behavior to explain above question concerns they ought have close relationship between social behavior and organizational strategic choice or organizational behavioral changing need.

On nowadays ecommerce business, or online shopping model, this kind of shopping model in global many young and old age consumers like to apply internet tool to choose any country sellers website stores in order to stay at home to buy any kinds of products from themselves webstores in global societies.

In fact, online shopping model had been popular for long time above to twenty years. Most of global sellers will make decision to design themselves webstores in order to attract global many online buyers to choose to buy their products from themselves webstores. So, it seems that social consumers purchase behaviors had been changed to online shopping from internet invention.

Hence, social consumers purchase behavioral changes may influence any organizations' strategies need to be changed from visiting shops purchase strategy model to online purchase strategy model, if the seller still concentrate on concentrate on considerate how to design itelf , but neglects to considerate how to design itself webstore, e.g. how to design attract product photos to put on itself webstore, how to arrange sale price information location to be putted on webstore and visa card payment location on itself webstore in order to let any one online buyer can feel very easier to buy itself any kinds of products from itself webstore. Then, its potential online buyers will be influenced to increase number when they can find this online seller itself any kinds of products photes and every kinds of product sale price information and visa card payment channel locations easily from itself webstore.

So, it implies that nowadays any one seller ought need to design one webstore to let any one online overseas and domestic consumers can have chance to click itself webstore to choose any one kind of product to buy conveniently when he/she does not hope to leave him/her home to go to shop, because nowadays social shopping behaviors had been influenced to change when internet invention, them it gives another online purchase method to replace visiting shops purchase method to global any one buyer in nowadays societies.

So, if nowadays any one seller still concentrate on how to design itself shop display in order to put any kinds of product on shelf in order to let any one visiting shop customer to find the kind of product to buy, but it neglects to change to choose to pursue another new technological shopping method, such as webstore purchase method in order to implement effective strategy to design the most right webstore as well as in order to attract global overseas and local consumers to find itself webstore easily from website and find its any one kind of product phots and sale price and visa card payment button in order to choose to buy itself any kinds of products in the short time. Consequently I believe that the seller will lose many customers from overseas and local when its other same or similar product sellers choose to design themselves webstores in order to let global any one product buyer can buy themselves any one kind of product when they can pay visa card to buy their products from them webstores conveniently when they stay at home habitly. Then, the seller will lose many global potential customers in long time.

On conclusion, in behavioral economic view, any consumer behavioral social changing, which will influence any in order to avoid customers number loses significantly . In future time, organizations need to make rapid decision in order to implement the most reasonable and the most useful strategy in order to avoid global potential customers number reduces or lose them in long time. So, social behavioral changing environment ought influence any global organizations need to decide how to change themselves strategies in order to avoid customers loses significantly in future time.

How and why human behavior may influence economic growth or recession?

May ourselves daily behaviors influence our global societial continue economic growth or recession? Do they have cause and effect close relationship between human behaviors and global economic growth or recession? I shall apply behavioral economic theory to analyze and explain whether ourselves daily behaviors and our global societial economic growth or recession which have close cause and effect relationship as below:

Every country itself economic development must depend on any business activities, otherwise, any kinds of business activities must need ourselves business activities or behaviors in order to achieve any business activities as well as achieve the country's overall economic development in macro view.

However, any country's overall business activites or behaviors which must depend on any kinds of individual businessmen, themselves employees daily working behavior or activity or performance in order to help them to attract or increase many clients number to acieve " earning profit" aim. So, it seems that any individual business, itself overall every department individual working behavior is one main factor to influence the company's overall business performance.

For agricultural fruit and meat food farming industry example, such as New Zealand is a farming main target industry country. It had had many New Zealanders were daily themselves own farming businesses for many years. Their farming businesses include growing fruit, sheep, cow, pig pork, meat etc. food sale business. If the New Zealand farmer owned a large size farming land, then he will choose either growing fruit or feeding sheeps, pigs, cows to be meat to to transport to New Zealand supermarkets to help them to sell to their farmers meet to New Zealanders in order to earn profit. Thus, if the New Zealand farmer owned large size of farming lands, then he needs to employ many farming employees (farming workers) to help him to carry on farming business daily tasks, e.g. picking up friuts, feeding pigs, cows, sheeps to eat food daily. These daily farming jobs are very important to influence this New Zealand farmer's meats or fruits sale number whether they can be easy or diffcult to sell in New Zealand supermarkets , if these farming workers can own encough farming knowledge or skill to know how to pick up fruits method and make judgement to know whether it is right time to pick up the kind of fruits from the trees , as well as know how feed this pigs, sheeps, cows to eat food in order to let they are better health. Consequently, their farming behaviors which can let these animals can provide the best taste and enough meat from these animals to let New Zealander to buy to eat from New Zealand any one supermarket. Even these New Zealand farming workers can know whether the kinds of fruits, e.g. oranges, apples, gapes etc. fruits whether they ought be picked up from the trees at the right time. Consequently, they can make judgement to decide to pick up any kinds of the best taste fruits to let any one New Zealander to buy to eat from any one supermarket in New Zealand. Otherwise, if they do not make judegement to know whether the kind of fruit ought not be picked up because they still need longer time to continue grow up to increase fruit size and better taste from the trees in order to let any one fruit buyer can feel better taste when they eat this kind of fruit later. If they can buy this kind of fruit to eat later, then this New Zealand farmer's his fruit buyers can buy the best taste of this kind of fruit to eat from an yone supermarket in New Zealand. Consequently, many New Zealand supermarkets will choose to buy any kinds of fruits from this farmer fruit supplier when they feel this farmer's fruits can provide more better taste fruits to compare other farmers' fruits.

Thus, due to New Zealand is one farming main income source country. It's any kinds of fruits and meats need to be export to overseas to sell , instead of local sale. It's GDP percent is very high to whole country 's overall income source. So, any one New Zealand farmer individual and any one farming worker individual working behavior will influence its economy whether it is influenced to grow or recession possible. Moreover, it also seems that farming workers' farming knowledge and skill will influence themselves farming daily activities to achieve the aim of the number of increase or decrease to any kinds of fruits whether they are better taste or the number of increase of decrease to any kinds of meats whether they are better taste to supply to any one New Zealand fruit or meat buyers to eat from any one New Zealand supermarket. So, it implies that any one New Zealand farming worker individual farming behavior may influence any kinds of fruits or any kinds of meat taste because they are transported to any one supermarket to sell in New Zealand.

Consequently, if New Zealans had many farmers can teach god farming knowledge and skill to let their any one farming workers know how to decide judgement to decide when it is right time to pick up any kinds of fruits from trees , or how to grow them on soil in order to let they can grow rapidly. Then, many different kinds of fruits can be provided to let any one New Zealanders can eat the best taste of fruits when their fruits are supplied to any one New Zealand supermarkets. Even, if they knew how to feed foods to pigs, cows, sheeps to eat daily. Then they can be more health and they can provide the best taste of meats to let any one New Zealanders can buy their meats from any one New Zealand supermarkets. Moreover, their fruits and meats can be transported to overseas to let any one country fruits or meats buyers can choose any kinds of New Zealand meats and fruits to buy to eat from themselves countries supermarkets. Then, many overseas fruit and meat buyers will perfer to choose New Zealand any kinds of fruits or meats to buy to compare other countries fruits or meats to buy when they go to any one local supermarkets.

On conclusion, it seems that New Zealand farming workers themselves farming behavior may influence their farming employers any kinds of fruits or meats sale number and income because their farming task behaviors must influence whether their fruits or meats taste are the better taste or worse taste to compare their other local farmers (the farmer competitors) whose fruits or meats taste. If tthe farmer's any one farming worker can be trained to learn how to know to feed animals skill and when is the most right time to pick up any kinds of fruits from trees or how to grow them on the soil methods. Due to these farming worker individual farming behavior may influence his different finds of fruits and meats sale number to be increase or decrease, so these any one

New Zealand farmer must need to depend on any one farming worker whose farming working methods, if their farming working behaviors can be the best to influence any kinds of fruits to grow rapid or any kinds of pigs, cows, sheeps animals grow up rapidly , then their sale number may be increase significantly and their taste can be improved to let any New Zealand or overseas meat or fruit buyer to buy to eat to feel from any one New Zealand or overseas supermarkets, then New Zealand's agriculture industry must be influenced to increase. In the world, any one fruit or meat buyer must choose to buy New Zealand's fruit and meat to eat in prefer to compare other countries' fruits and meats. So, New Zealand's GDP may be influenced to raise from any one New Zealand farming worker individual farming working behaviors.

www.ingramcontent.com/pod-product-compliance
Ingram Content Group UK Ltd.
Pitfield, Milton Keynes, MK11 3LW, UK
UKHW040556210726
13854UKWH00007B/509

9 798888 831502